FRUSTRATIONS WITH MATH

FRUSTRATIONS WITH

MATH

JERRY ORTNER

ReadersMagnet, LLC

Frustrations With Math

Published in the United States of America
ISBN Paperback: 978-1-951775-31-5
ISBN eBook: 978-1-951775-32-2

ReadersMagnet, LLC
10620 Treena Street, Suite 230 | San Diego, California, 92131 USA
1.619.354.2643 | www.readersmagnet.com

Cover design by Ericka Obando
Interior design by Shemaryl Tampus

CONTENTS

DAY 1

ADDING MINUTES TO HOURS

As I began my road trip, what caught my attention as my fully packed Chrysler Sebring convertible tried to enter I-95 and head south, from Marsh Road in Wilmington, Delaware, was an automobile sitting on a fence. Lo and behold, the driver not only skidded on the slick entrance way to I-95 but had successfully removed himself from the auto and called for assistance. While he was patiently waiting for the flatbed truck to somehow remove the auto, I sat waiting for the festivities to begin.

Did I say it took long?

Oh no, after just 10 minutes of waiting, I changed gears, tooted my horn to let the one car behind me back up, and I navigated the melting snow bank and off I went. As I left, so did the tow truck, and everyone was relieved to begin their promenade heading south on I-95.

Not I. Needless to say, my split second decision to try another way was reversed when the roadway cleared and I meekly retraced my steps and plodded back on track to my ultimate destination of the Autotrain Station in Lawton, Virginia.

Driving down the interstate, my only objective was to get to the Autotrain Center before 2 PM. It was 11 AM and the trip ticket from MapQuest(c) said it would take about two hours, twenty-five minutes. When one plans a trip, who thinks of total minutes. Just hours would be easier. As a math teacher, I had to begin calculating 10 minutes added to two hours, twenty-five minutes gave me only a twenty-five minute cushion.

How did I calculate that you asked? 11 AM to 2 PM gave me three hours. That's 180 minutes (remember: 60 minutes to an hour). Sixty times three equal 180 minutes. From 180 minutes, subtract two hours, or 120 minutes, and you have 60 minutes. [For those of you thinking that it is much easier to play with hours, then it is 3 – 2 or 1 hour which is 60 minutes.]

Now comes the tricky part. Subtract the 25 minutes from 60 minutes and I have 35 minutes to spare. But I just lost 10 minutes being entertained by a tow truck removing a car on a fence. So an extra 25 minutes is fantastic.

Once out of Delaware, rolling down I-95, Maryland is the next state. Everything was going smoothly and fast. Twenty-five minutes to spare and that should be enough.

Ouch. An overhead sign ahead says the tunnel traffic is delayed a minimum of five minutes due to the closure of the inner tunnel. Five minutes from 25 minutes makes 20 minutes maximum to the Autotrain.

Do I skirt the minimum five-minute delay or do I take the outer loop? Trip ticket said to use I-95 all the way so I can't deviate from the plan. Just hope the tunnel delay is not more than five minutes. I paid my toll and proceeded to take the left lane of the western tunnel since trucks would be on my right.

I guessed wrong. Trucks and cars whistled by me on my right as I looked at the clock. The red lights of the cars in front of me didn't lend much encouragement. Do I dare cross the solid line in that tunnel? The car behind me did but I would be the lucky one and have a state trooper just waiting for me to cross the line. I persevered and luckily only a five-minute delay was incurred.

Down to twenty minutes and counting. I noticed state troopers pulling over fast moving vehicles, so I maintained the 65 mph speed limit but don't exceed that limit. Cruise control would be useful but not every car has it. Mine doesn't. Just a few years to soon for cruise control or not new enough. You be the judge.

Where's the outer loop around Washington, DC? Can't come soon enough. And it did. I'm sailing along without any troubles. Keep your fingers crossed as I head west and into Virginia. No radar detectors! Put it away. Just hope there are no more delays.

I was lucky. I came down the I-95 ramp at exit 163 and turned left. Just a few yards to my left was the Autotrain. I pulled into the Autotrain entrance and looked ahead. Cars, cars, and more cars were lined up waiting their turn to pass through the initial checkpoint. Only twenty minutes to spare. Was my car going to make the 2 PM deadline? The single line moved slowly. Five minutes elapsed. It was 1:45 PM. Please, please move faster. Now ten minutes until the gong sounded. I counted 10 cars in front of me and lots in back. Close my eyes and wish. Now open them and glance at the line. Only three cars to go with five minutes to spare.

Did I make it, you asked? Check out Day 2 of the road trip through math.

DAY 2

PYTHAGOREAN THEOREM
(Right Triangle You Say)

YES, I DID MAKE IT WITH one minute, twelve seconds to spare. Five more cars came in behind me as the clock struck 2 PM.

Incidentally, as a refresher from Day 1, one minute, twelve seconds translates into 72 seconds. That is longer than a 60-second Super Bowl commercial which costs over two million dollars. Another way to say that is, seven ten-second commercials for half a million dollars each. Can you write them apples? How about starting to count! One, two, three, four, five – oh, I give up!!

On to Day 2 of my road trip. Here's the scoop on the workings of an autotrain. It is much like railroads today. An all-passenger train carrying automobiles from here to there. The here is Lawton, Virginia and the there is Sanford, Florida. All in the span of sixteen and one-half hours, give or take. My guess, it is the give. It will take more time to reach the destination in Florida on this particular evening/morning.

If you agreed with me, you're incorrect. We arrived one-half hour late. Playing with those minutes again, it adds up to seventeen hours. That's on a train. Lots of moving around, shuffling the body as the train rocks side to side, but lots of fun. How many minutes?

During check-in, dinnertime choices had to made. I selected 5 PM as my dinnertime. Only three choices and 7 PM was sold out. That left 5 PM and 9 PM. Nine PM was too late to eat, digest and relax. Five PM was the magic hour. Since I didn't stop for lunch, I was hungry by 3 PM, so 5 PM was it. Train service and food was superb. The waitstaff was pleasant and eager to be helpful. Three main entrees were presented. All three were great as my table guests attested to the fact that, "Hey, this ain't bad!"

One of the dinner guests at my table was Cliff. He was a retired civil engineer and had recently purchased a marina on the shores of the Chesapeake Bay. Another guest at dinner was an author of cookbooks. Her specialty was "all kinds of beans". I didn't pursue the topic as I was too frightened to ask, "What type of bean? Green, yellow, kidney, soy, ..." you get my drift.

Kay, the author, asked what I did and I explained my road trip to relieve myself of math frustrations. Kay thought it was a fantastic idea but couldn't seem to see how I could write about math problems and keep it interesting. With a little coaxing, I said, "Over "x" number of years teaching or explaining math, I should know some phobias."

Cliff chimed in and said, "As an engineer, I know my math. But what drives me crazy is all the yacht owners. They don't know how to get from one location to another without getting lost or using that compass."

My interest peaked and I inquired why would anyone who owned a yacht get lost?

"Because they never learned about the right triangle relationship," remarked Cliff.

"The right triangle relationship?" asked Kay.

"Yes," explained Cliff. "No one can figure out how to get from point A to point B without going to point C."

We all listened alertly as Cliff did the Pythagorean Theorem hypothesis in lay peoples' terms. I felt his explanation was quite good. In fact, it went to the top of my list of explanations of how

$a^2 + b^2 = c^2$ where a and b are the legs of the right triangle and c, which is called the hypotenuse, is the side opposite the right angle.

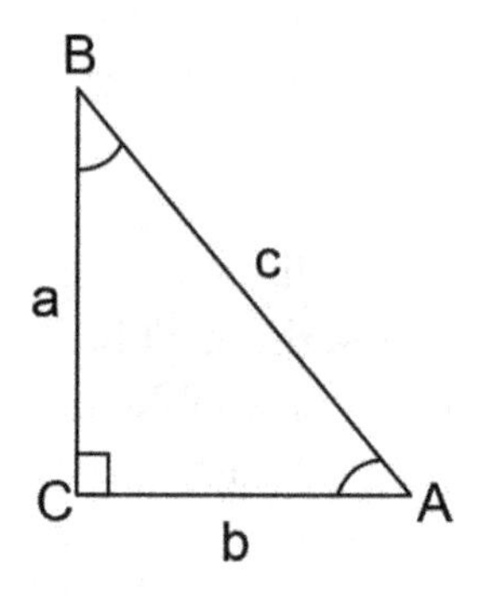

$$a^2 + b^2 = c^2$$
Angle C = 90°
($\angle C = 90°$)

measure of angle A + measure of angle B = 90°

Cliff explained that, in theory, to get from point A to point B anywhere on this globe, select another point, C, so that the angle formed at C is 90° to the "rays" of ∠A and ∠B. Use that reference and the numbers will work. I inquired about the numbers. Cliff used the 3-4-5 ratio to summarize his numbers. If the distance between A and C is 3 miles and the distance between B and C is 4 miles, which total 7 miles altogether, then the distance A to B is only 5 miles. That's two miles shorter. For yacht owners, two miles can be forever if winds or fuel are shortcomings.

My only comment was directed towards golf, which neither of my dinner companions played. It centered about the green at one hole at a notable golf course in Arizona. There was a sand trap in the center of the green. Should I try to get close to the pin "hole" or just get it onto the green? You guessed it. I hit a great shot but it was on the wrong side of the green. Did that right angle theorem come into play? Yes, it did. I picked a point at right angles to my position so I could get par in two shots.

On that day long ago, someone in my foursome landed in the trap, pitched it in from there and walked off with a birdie. Golfers are always lucky. Or so it seems.

The Autotrain ride produced no other major surprises or ideas, so off I traveled to Southwestern Florida.

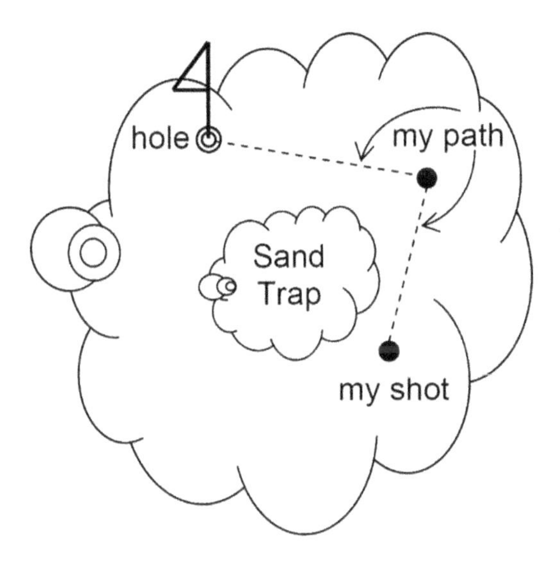

DAY 3

RESTAURANT TIPPING

(special thanks to Patty Hamway of Marco Island, FL)

ALL WENT WELL ON MY JOURNEY to Southwestern Florida. I was heading south of Naples, specifically, near Marco Island. The digs are great. Beautiful views of a golf course and numerous lagoons. My day was filled with getting the "lay of the land", and finding out what activities were available for an average duffer.

First things first. My taste buds needed a filling. I ventured out and found a very nice restaurant that was open 24 hours, 24/7 we call it. An omelet with home fries, juice and coffee sounded good. The cost was $9.18, which included tax. What tip should I leave? I read that anywhere between 15% and 20% is appropriate. Eighteen percent is sometimes added directly to the bill so there is no thinking involved. Since 18% is close to 20% let's use that number to calculate the tip. Ten percent of $9.18 is 91.8¢ and double that is $1.836. So let's see, $9.18 plus $1.83 equals $11.01. Leave $11 and walk away. Now when the numbers get larger, it is so much easier.

Let us use the example of a restaurant bill of $50. Eighteen percent of that is $9. Just take 10% of $50 ($5) and 2% of $50 ($1), so 20% of $50 is $10 minus the $1 gives you $9.

Restaurant tipping is just like a sales tax. Some states, and even cities, charge a fractional percent, $7\frac{1}{2}\%$ or $8\frac{1}{4}\%$. You do need a calculator for that problem. Unfortunately, my calculator just went BLANK. Yep, the batteries died. Did I have any AA batteries in my pocket? No such luck. If I have to buy batteries then I need them before I pay the bill so the calculation can be completed and I know what I owe. Yes, the store will gladly figure it for me but, being a frustrated math teacher, I like to know what my costs will be.

Haven't you gone shopping (food, clothes, gadgets, etc) and calculated, in your head, how much you spent? Try it the next time when getting numerous items and you will be amazed.

Let's go back to the restaurant and what to give for a tip.

We used 18%, the current industry standard. What if the service is better than average? Give the server 20%. An easy calculation. Just multiple the bill by 1/5 or 0.2 and the number will appear.

If the service is below average, give 15%, which is harder to figure but it works. Take 10% of the bill and add another 1/2 of the 10% and, presto, you've got the answer!

Reviewing quickly: take a restaurant bill of $40. A 20% tip would be $8 and a 15% tip would be $6.

Dinner for 3	
entrée	9.95
entrée	10.70
entrée	10.50
drink	1.25
drink	1.25
drink	1.25
dessert	2.50
dessert	2.25
dessert	1.75
total	41.40
41.40 × .15 = 6.21	
41.40 × .20 = 8.28	

Buffet for 2 adults and 2 children	
buffet	7.30
buffet	7.30
buffet	4.50
buffet	4.50
total	23.60
23.60 × .15 = 3.54	
23.60 × .20 = 4.72	

Lunch for 2	
sandwich	3.95
sandwich	4.55
drink	1.25
drink	1.25
dessert	2.50
dessert	2.25
total	15.75
15.75 × .15 = 2.36	
5.75 × .20 = 3.15	

I'm curious what Day 4 will bring. Turn the page and get out the calculator.

DAY 4

OUTLET SHOPPING

Learning what a discount truly means.
(again thanks to Patty Hamway)

THE BEAUTIFUL WEATHER TURNED BLEAK AND overcast on Day 4. I tried to rationalize playing with my sticks (slang for golf clubs) but the grey sky, blustery winds, and drizzle meant only one thing: GO SHOPPING at the outlets!

Everything in the outlets had sale, sale, sale in the windows. Did that mean a true sale or just a gimmick to get you into the store? Both assumptions were correct.

Not only where there true discounts with 50% off, but the clearance tables were fantastic. I forgot my calculator to figure how much I could save, so I "adlibbed" it. Wouldn't you guess, my hunch was right. Who needs another pair of khaki shorts? Or better yet, black jeans? Not I, so I moved around the corner to what looked like a fantastic deal.

Buy the first pair of shoes and get another pair, of equal or lesser value, for 50% off. What a deal!! Maybe I can find a size 12M golf shoe, in black, and get something else for 1/2 off. I liked the foot joy of golf shoes and it came in black, size 12 medium. Only trouble was that it sold for $50, and that was the sale price. Not a bargain for an outlet store. Ok, I gave in and then thought, what can I buy for 1/2 off. I glanced at the clearance rack and found the only 12M box in the selections. Trouble was the box only had one right shoe. It was a white fancy sneaker with a very popular name, Tony Hilfiger. I know that name and exclaimed, "Wow! Fantastic!" I looked more closely at the box and it said, "Ask for the left shoe." Moving rapidly to the counter, clutching the black pair of golf shoes in one hand and the single TH sneaker in the other, I inquired about the left one. The smiling clerk, without batting an eye, reached into a drawer below the counter and, presto, the left size 12M, TH, sneaker appeared. My excitement was euphoric. I found two pairs of footwear on sale.

Trouble began when I could not figure out which to buy first or second. The sneakers cost $40, so I had a dilemma. Which pair to buy first?

With no calculator, nor a pencil, in hand, I began to figure out, mentally, how I should begin. Do I buy the $50 golf shoes first, and get 50% off the sneakers, or vice versa? What a bargain, I thought?

Now I must work out the not so tricky problem of which pair of shoes to buy first. To maximize my savings, I need to be sure to select the right order of purchase.

First, I tried the $50 golf shoes and then the $40 sneakers. Let's calculate: $50 plus 1/2 of $40, which is $20, rings up $70 on the register. If I reverse the order, $40 for the sneakers plus 1/2 off the $50 golf shoes (that's $25), the math adds up to $65.

I found the correct order of purchase. Buy the cheaper item first and then the other. The clerk looked in amazement as she rang up the order and said, "That will be $70 please." I cleared my throat, coughed again, and said, "What?" She politely said, "$70".

I was dumbfounded. How could I, a math teacher for so many years, get SOMETHING wrong. My adrenaline began to rise as I inquired calmly, "How can that be?" The smiling clerk knew she had a winner (loser in my book). The sign on the register indicated that the 1/2 price on the second pair must be of the same or lesser price of the first.

I had been had! Two hads in a sentence. Beyond belief. My ego got shot down but I was happy to get another pair of golf shoes and a pair of TH sneakers for $70. My credit card company loves me spending at the outlets.

However, my wife, who instigated the outlet shopping, whispered, "Oh, well", and raised her hands and she, of course, was smiling too. What's that called in chess ... CHECKMATE! I lose!

DAY 5

READ THE STORY PROBLEM CORRECTLY

THE EXCITEMENT FROM DAY 4 WAS tempered by my shortsightedness in NOT READING the sign at the outlet. How true. If I just stopped and read the posted notice, my deliberations would have been unnecessary.

Students, when studying math, seem to be overwhelmed by story problems. What did the question ask? How many? Which one? How long? etc, etc, etc.

My favorite example that would amuse students at every level is the following story:

It takes four hours to travel from Denver to Colorado Springs. Henry left Denver at 9 AM and after awhile, checked his watch. It was 11:25 AM. How much longer must he drive to reach Colorado Springs in four hours?

$$\begin{array}{r} 11:25 \\ -9:00 \\ \hline 2:25 \end{array}$$

This translates into 2 hours 25 minutes.

A lot of students would answer 2 hours 25 minutes. Go back and re-read the question. How much longer must he drive to get to Colorado Springs? Not how long has he driven.

So after getting 2 hours 25 minutes, that answer must be subtracted from four hours.

$$\begin{array}{r} 4:00 \\ \underline{-2:25} \end{array}$$

Need to change one hour to 60 minutes, then proceed.

$$\begin{array}{r} 3:60 \\ \underline{-2:25} \\ 1:35 \end{array}$$

The correct answer is 1 hour 35 minutes.
Try this one: Divide 3472 by 16. What is the remainder only?

$$\begin{array}{r} 217 \\ 16\overline{)\,3472} \\ \underline{-32} \\ 27 \\ \underline{-16} \\ 112 \\ \underline{-112} \\ 0 \end{array}$$

The answer is <u>zero</u>. Were you right?

Try this one:

Given that 2 cups equal 1 pint and 2 pints equal 1 quart, how many cups are there in 7 quarts?

$$7\ \cancel{\text{quarts}} \times \frac{2\ \cancel{\text{pints}}}{2\ \cancel{\text{quart}}} \times \frac{2\ \text{cups}}{1\ \cancel{\text{pint}}} =$$

$$7 \times 2 \times 2 \text{ cups} = 28 \text{ cups}$$

As long as you set up the proportion like that above, where the units of measure cancel, you will be okay.

One more for good measure:

16 is 40% of what number?

Solving this proportion requires knowing how to convert a percent to a fraction with 100 as the denominator and then using this formula:

$$\frac{\text{Part}}{\text{Whole}} = \frac{\text{Percent}}{100} \quad \longleftarrow \text{Remember this!!}$$

$$40\% = \frac{40}{100}$$

$$\frac{16}{N} = \frac{40}{100} \text{ cross multiply } \frac{a}{b} = \frac{c}{d} \Rightarrow ad = bc$$

$$16(100) = 40N$$

$$\frac{1600}{40} = \frac{40N}{40}$$

$$40 = N$$

16 is 40% of ***40***.

> Remember, read the story problem several times to know what is asked. Then check your answer to see if it is reasonable

We'll discuss this later on Day 18.

DAY 6

BASE 10 NUMBERS

IN THE 60'S DECADE, "NEW MATH" revolved around base numbers such as base two, base five, and the one everyone uses, base ten. When one counts in base ten, after the first nine single digits, we move one place further to the left and create another placeholder. When the number 99 appears, we again move one further place to the left. This keeps repeating over and over through the thousands, million, billions, etc.

What everyone tries to fathom is how many zeros do I need to write the number. In base 10, a simple way to write the number is as follows:

$10^7, 10^6, 10^5, 10^4, 10^3, 10^2, 10^1, 10^0$

Remember:

$10^0 = 1$
$10^1 = 10$
$10^2 = 10 \times 10 = 100$
$10^3 = 10 \times 10 \times 10 = 1000$

Note: Any number, except zero, raised to the zero power is 1.

Try these:

1. 10^4 = __ × __ × __ × __ = ______
2. 10^5 = __ × __ × __ × __ × __ = ______
3. 10^6 = __ × __ × __ × __ × __ × ___ = ______
4. 10^7 = __ × __ × __ × __ × __ × ___ × ___ = ______
5. 10^8 = __ × __ × __ × __ × __ × ___ × ___ × ___ = ______

The key is place value:

ten million	million	hundred thousand	ten thou-sand	thousand	hundred	tens	ones

Try some problems. Write in words:

6. 6421______________________________
7. 372,612____________________________
8. 4,506,075__________________________
9. 37,402_____________________________
10. 672,345,981________________________

And the reverse. Write in numbers:

11. sixty-three million, two hundred twelve thousand, five hundred fifty-seven
12. eighty-nine thousand, seven hundred thirty- nine.

13. nine hundred twenty-one thousand, six hundred seventy-three
14. two hundred thirty-four million, five hundred sixty-nine thousand, four
15. four thousand seventy

In conclusion, my primary objective was to alert you to place values and how the decimal number system works. Everyday occurrences, such as writing checks, can be overwhelming and burdensome if you are unfamiliar with place values. One never knows when a check needs to be written for ten dollars, one hundred ten dollars or, even, one thousand ten dollars.

My motto is "BE PREPARED!"

DAY 7

HOW TO CHEAT WHEN ADDING OR SUBTRACTING FRACTIONS

LET'S LOOK AT THE METHOD EVERYONE was taught but somehow forgot: $\frac{2}{3}+\frac{4}{7}$

$$\frac{2}{3}\times\frac{7}{7\,\boxed{B}}=\frac{14\,\boxed{D}}{21\,\boxed{A}}$$

$$\frac{4}{7}\times\frac{3}{3\,\boxed{C}}=\frac{12\,\boxed{E}}{21\,\boxed{A}}$$

$$\frac{26\,\boxed{F}}{21\,\boxed{G}}$$

A	Common denominator for 3 and 7 is their product: $(3 \times 7 = 21)$.
B	Divide: $21 \div 3 = 7$
C	Divide: $21 \div 7 = 3$
D	Multiply: $\boxed{B}\ 7 \times 2 = 14\ \boxed{D}$
E	Multiply: $\boxed{C}\ 3 \times 4 = 12\ \boxed{E}$
F	Add: $\boxed{D}\ 14 + \boxed{E}\ 12 = 26\ \boxed{F}$
G	Keep the same denominator ($\boxed{A}$):21 $\boxed{G}$

Now we have a solution, $\frac{26\,\boxed{F}}{21\,\boxed{G}}$ reducing, $\frac{26\,\boxed{F}}{21\,\boxed{G}}=1\frac{5}{21}$, if necessary, and we're done!

Now let's cheat!

$$\frac{A}{B}+\frac{C}{D}=\frac{(A)(D)+(B)(C)}{(B)(D)}=\frac{AD+BC}{BD}$$

Using the above example:

$$\frac{2}{3}+\frac{4}{7}=\frac{(2)(7)+(3)(4)}{21}=\frac{14+12}{21}=\frac{26}{21}=1\frac{5}{21}$$

It is a lot faster and it always works!

$$\frac{5}{8}-\frac{1}{16}=\frac{(5)(16)-(1)(8)}{128}=\frac{80-8}{128}=\frac{72}{128}\text{, which}$$

$$\text{reduces by 8, } \frac{72}{128}\frac{\div 8}{\div 8}=\frac{9}{16}$$

Let's try mixed numbers. Change the mixed numbers to improper fractions.

$$4\frac{1}{3}+5\frac{3}{5}+7\frac{1}{4}=\frac{13}{3}+\frac{28}{5}+\frac{29}{4}$$

Add the first two fractions:

$$\frac{13}{3}+\frac{28}{5}=\frac{(13)(5)+(28)(3)}{15}=\frac{149}{15}$$

Then add to the third fraction:

$$\frac{149}{15}+\frac{29}{4}=\frac{(149)(4)+(29)(15)}{60}=\frac{596+421}{60}=$$

$$\frac{1017}{60}=16\frac{19}{20}$$

Change the problem just a little: $4\frac{1}{3}+5\frac{3}{5}-7\frac{1}{4}=$

$$\left(\frac{13}{3}+\frac{28}{5}\right)-\frac{29}{4}$$

Add the first two fractions:

$$\frac{13}{3}+\frac{28}{5}=\frac{(13)(5)+(28)(3)}{15}=\frac{149}{15}$$

Then subtract the third fraction:

$$\frac{149}{15}-\frac{29}{4}=\frac{(149)(4)-(29)(15)}{60}=\frac{596-421}{60}=$$

$$\frac{175}{60}=\frac{35}{12}=2\frac{11}{12}$$

It's even good for ALGEBRA!

$$\left(\frac{a}{3}+\frac{7b}{4}\right)-\frac{3c}{5}=\left(\frac{4a+21b}{12}\right)-\frac{3c}{5}=$$

$$\boxed{\frac{20a+105b-36c}{60}}$$ Does not reduce!

My thanks to Patty Hamway of Marco Island for the fractions suggestion. Every once in a while it is nice to be able to play with fractions and KNOW that the numerator divided by the denominator can by simplistic.

Keep on smiling as math can be enjoyable. Patty did recommend that knowing her multiplication tables made her job much easier. Isn't it strange that individuals, knowing their multiplication tables, can enjoy the rigors of math. Others, who rely on a calculator, might say "Who cares!"

If the calculator fails, due to dead batteries, who is to blame? It is the elementary math teacher or math program that never stressed the importance of knowing the multiplication tables, of course!

Make a 10 × 10 grid and fill in the spaces. The outside is easy, When you get out to the far end of the table, say, sevens, eights and nines, it becomes harder. Work the grid until all 100 answers can be done in 60 seconds.

At that time, you can then say you're a smiling, happy, successful mathematician!

P.S.: Make sure you cover up the answer grid.

	1	2	3	4	5	6	7	8	9	10
1										
2										
3										
4										
5										
6										
7										
8										
9										
10										

	1	2	3	4	5	6	7	8	9	10
1	1	2	3	4	5	6	7	8	9	10
2	2	4	6	8	10	12	14	16	18	20
3	3	6	9	12	15	18	21	24	27	30
4	4	8	12	16	20	24	28	32	36	40
5	5	10	15	20	25	30	35	40	45	50
6	6	12	18	24	30	36	42	48	54	60
7	7	14	21	28	35	42	49	56	63	70
8	8	16	24	32	40	48	56	64	72	80
9	9	18	27	36	45	54	63	72	81	90
10	10	20	30	40	50	60	70	80	90	100

DID YOU FINISH IN LESS THAN 60 SECONDS?? If not, do the exercise again and again until you can complete the chart in less than 60 seconds.

DAY 8

PRIME NUMBERS

NOW WE ARE SEEING WHAT MATH is all about. As I hit Day 8 on my road trip, prime numbers are on today's agenda. Knowing primes will help you to factor. Math has crescendo effect. It keeps on building and building and building.

What made me select primes?

When I inquired from someone, "What frustrated you most in math?" The answer was factoring in algebra. In order to factor, a topic which will be discussed later, one must understand the prime numbers.

By definition, a prime number has only itself and one as its product. It is this understanding of primes that enables students to factor. You have probably been asked sometime during your math experience: "What two numbers when multiplied, give you 20?"

Your answer might be 1 and 20. Why not? It is a good choice. However, several other combinations work: 2 × 10 and 4 × 5, also written as: (2)(10) and (4)(5). That gives us three sets of factors for 20.

How does factoring relate to prime numbers? Start with the chart of all one- and two-digit numbers, exclusive of 1. The first number is 2 and it is prime. The next number is 3, which is also prime. We skip 4 since (1)(4) and (2)(2) equals 4. Every even number after 2 is ***not*** prime. Then proceed to 5, which is prime. 7 is also prime. Why not 9? Because its factors are (1)(9) and (3)(3).

Therefore, the only single digit primes are: 2, 3, 5, and 7.

Two-digit primes:

11, 13, 17, and 19 are all primes.

15 has factors of 1 × 15 and 3 × 5. Now all multiples of 5 are not prime.

21 [(7)(3) and (1)(21)], 25 [(5)(5) and (1)(25)] and 27 [(3)(9) and (1)(27)] are not primes.

23 and 29 are primes.

31, 37, 41, 43, and 47 are primes.

Let's recap all primes below 50.

2, 3, 5, 7, 11, 13, 17, 19, 23, 29, 31, 37, 41,43, and 47
(fifteen numbers in all so far).

Moving along the prime road, 53 and 59 are primes (51 has 17 and 3 as factors and 57 has 19 and 3 as factors). Just a note: when the sum of the digits are a multiple of 3, it will not be prime: 51 adds up to 6 (5 + 1) which is 2 × 3 and 57 adds up to 12 (5 + 7) which is 4 × 3.

In the sixties, 61 and 67 are primes. For the seventies, there is 71, 73, and 79. 83 and 89 are the eighties primes and, finally, only 97 in the nineties is prime.

Now comes the fun. All other numbers, seventy-five numbers less than 100, are composite. By definition, you can say that a composite number has more factors than itself and one.

Examine 48.

It has five pairs of factors: (1)(48), (2)(24), (3)(16), (4)(12), and (6)(8).

Reviewing all prime numbers less than 100 are:

2, 3, 5, 7, 11, 13, 17, 19, 23, 29,
31, 37, 41, 43, 47, 53, 59, 61,67,
71, 73, 79, 83, 89, 97

So to summarize with a chart:

	2	3	~~4~~	5	~~6~~	7	~~8~~	~~9~~	~~10~~
11	~~12~~	13	~~14~~	~~15~~	~~16~~	17	~~18~~	19	~~20~~
~~21~~	~~22~~	23	~~24~~	~~25~~	~~26~~	~~27~~	~~28~~	29	~~30~~
31	~~32~~	~~33~~	~~34~~	~~35~~	~~36~~	37	~~38~~	~~39~~	~~40~~
41	~~42~~	43	~~44~~	~~45~~	~~46~~	47	~~48~~	~~49~~	~~50~~
~~51~~	~~52~~	53	~~54~~	~~55~~	~~56~~	~~57~~	~~58~~	59	~~60~~
61	~~62~~	~~63~~	~~64~~	~~65~~	~~66~~	67	~~68~~	~~69~~	~~70~~
71	~~72~~	73	~~74~~	~~75~~	~~76~~	~~77~~	~~78~~	79	~~80~~
~~81~~	~~82~~	83	~~84~~	~~85~~	~~86~~	~~87~~	~~88~~	89	~~90~~
~~91~~	~~92~~	~~93~~	~~94~~	~~95~~	~~96~~	97	~~98~~	~~99~~	~~100~~

More on the topic of factoring on Day 13.

DAY 9

BUYING VARIOUS OUTFITS

Combinations

NOW EVERYONE IS WONDERING HOW THE notion of combinations enters a frustrated mathematician's mind. Very easily.

Picture yourself walking into a clothing store. You spot a fantastic colored blouse. The color is Moroccan blue. Now, what color shorts? Khaki. So you have a new outfit. Blue blouse with khaki shorts. Bravo!

Remember on Day 4 we spotted a bargain at the outlets? Again we spotted this opportunity to buy an outfit, blouse and shorts, for half price.

Let's see how many different combinations you can show when you have two different colored blouses, blue and green, with two pairs of shorts, khaki and white.

One set could be the blue blouse with the khaki shorts, your first purchase. Another outfit would be the green blouse and white shorts. If you interchange the tops, you would get two more outfits, blue blouse with white shorts and green blouse with khaki shorts. Four different outfits for the price of two.

Visually,

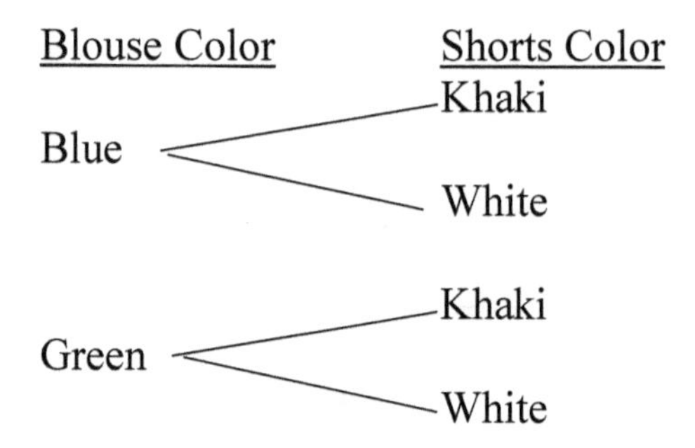

If you purchased a pair of khaki slacks, how many different combinations could you find?

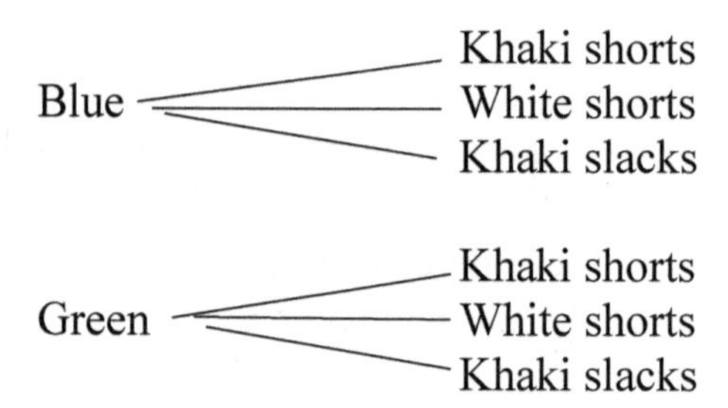

Two blouses times three shorts/slacks (2 × 3) gives you 6 different outfits!

What if you had four blouses, 2 shorts, and 1 pair of slacks? What would be the new combination?

Four blouses times 3 shorts/slacks would be 12! Did you get it right?

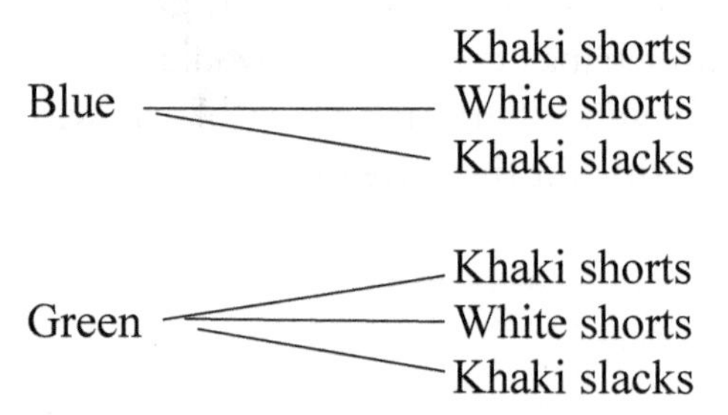

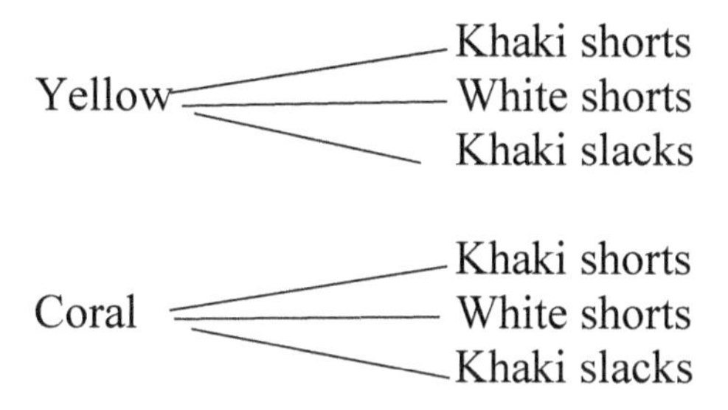

Three and one-half outfits, cost-wise, gives you 12 various combinations. Something to remember when you are planning your new wardrobe for the coming season.

Have fun shopping and keep in mind combos. We math teachers call it combinations.

DAY 10

OLLIE HAD A HEADACHE OVER ALGEBRA

"Trig"

MANY, MANY YEARS AGO, MY HIGH school math teacher, Roland Carney, used the title of Day 10, Ollie Had A Headache Over Algebra (OHAHOA) to help us remember the functions for basic trigonometry. Mr. Carney was the only high school math instructor in a small school system in western New York state.

Having the same math teacher for four years can be either good or bad. My experience, as you may have guessed, was great. He and I clicked immediately. It was his influence, among others, which enabled me to excel in math and gave me the desire to teach and help students achieve a sound math background.

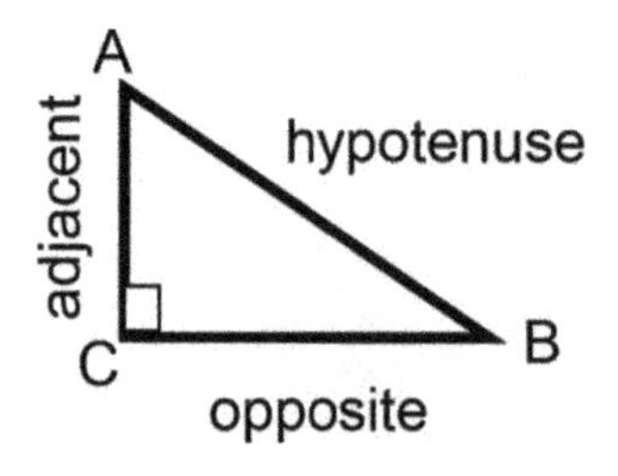

OHAHOA is an anachronism for right triangle trigonometry (trig). Trig is usually taken after completion of Algebra II. It (meaning OHAHOA) refers to the sides of a right triangle. I think a diagram would be appropriate now to explain this "cheating" method.

Given that angle C equals 90°, the sum of the other two angles must equal 90°.

Trig has three basic functions: Sine (sin), Cosine (cos), and Tangent (tan).

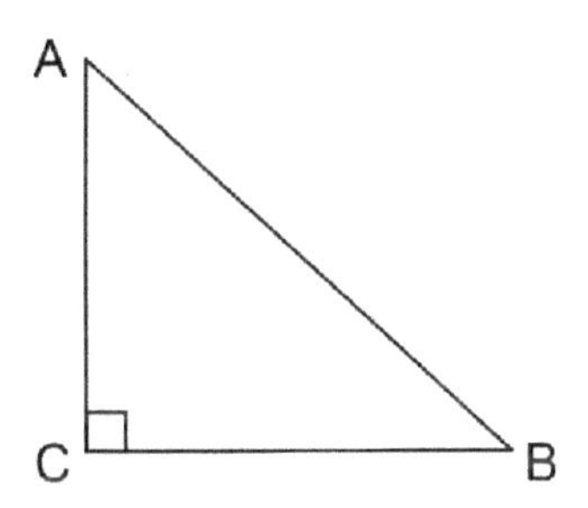

$$\sin\angle = \frac{\boxed{\text{O}}\text{pposite side}}{\boxed{\text{H}}\text{ypotenuse}}$$

$$\cos\angle = \frac{\boxed{\text{A}}\text{djacent side}}{\boxed{\text{H}}\text{ypotenuse}}$$

$$\tan\angle = \frac{\boxed{\text{O}}\text{pposite side}}{\boxed{\text{A}}\text{djacent side}}$$

Problem 1: Find the measure of angles A and B.

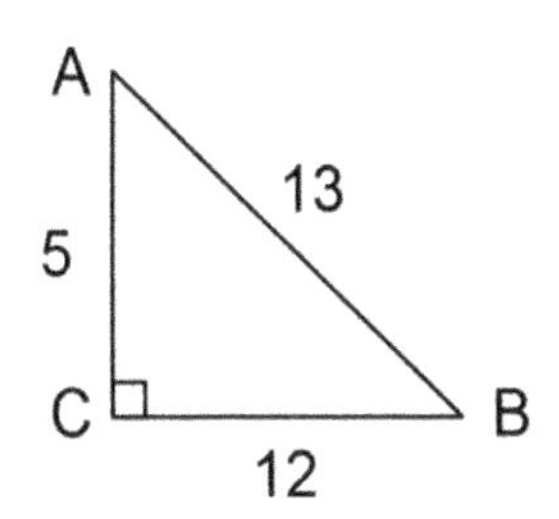

$$\sin \angle A = \frac{12}{13} = 53° \; 34'$$

$$\sin \angle B = \frac{5}{13} = \_\_\_\_\_\_$$

It is a lot easier to use the Pythagorean Theorem and the anachronism OHAHOA to find the lengths of the sides of a right triangle relationship.

Problem 2: Given AB side of 5 and BC side of 4, find the following:

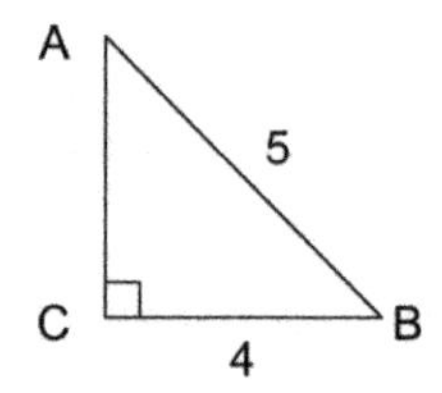

length of $\overline{AC}$
m $\angle A$
m $\angle B$

$$(\overline{AC})^2 + (\overline{BC})^2 = (\overline{AB})^2$$
$$(\overline{AC})^2 + 4^2 = 5^2$$
$$(\overline{AC})^2 + 16 = 25$$
$$-16 \quad -16$$
$$(\overline{AC})^2 = 9$$
$$(\overline{AC}) = 3$$

a. m $\angle A = \sin \frac{4}{5} = \_\_\_\_\_\_$ b. m $\angle B = \cos \frac{4}{5} = \_\_\_\_\_\_$

A calculator with trigonometry functions would be of great help here.

DAY 11

FRUSTRATIONS OF A STUDENT WITH HIS MATH INSTRUCTOR

Ike, a college graduate from Panacea, FL, was asked by me, "What was your most frustrating experience with math?"

His response was: "I was able to solve problems using a method different than the professor. However, the professor always deducted a few points from every answer that, even though correct, did not follow his method of instruction.

I agree with Ike. Many times a problem can be solved in several ways. Why penalize the student if their work in progress leads to an accurate solution.

My case in point:

A student received these grades: 84, 93, 87, and 98.
What must s/he get on the 5th exam to have a 91 average?

Solution #1:

Add 84 + 93 + 87 + 98. Their sum is 362.
Now 91 × 5 = 455. Subtract 362 from 455.
The answer is 93 or better.

Solution #2:

$$84 + 93 + 87 + 98 + N = 91(5)$$

$$\begin{array}{rcl} 362 + N & = & 455 \\ -362 & & -362 \\ \hline & N = 93 & \end{array}$$

Solution #3:

Average of 84 and 98 is 91. $\frac{84+98}{2} = 91$

Average of 87 and 93 is 90. $\frac{87+93}{2} = 90$

> 84 + 98 gets me 91, which I want. But 87 + 93 leaves me 90. I need 2 more points than 91 to get a 91 average. Therefore, a 93 or better score on the fifth exam is necessary.

There are other thought processes which probably result in the correct solution. Why penalize a student for not solving the way it was shown? Isn't education a learning experience? Trial and Error. Let the student be creative. Then we can have a variety of opinions and that is what education is all about.

Thanks, Ike, for your frustration.

DAY 12

DISTANCE, RATE, TIME

MY ROAD TRIP HAS TAKEN ME, by means of the Autotrain, to Florida. Once in Florida, I began calculating my distance covered every hour while driving from Sanford, Florida to Naples, from Naples to Shell Point, Florida, and from Shell Point to Hilton Head Island, South Carolina.

Travel time varies depending upon conditions. Heavy traffic, road repair, and speed limits are three such impediments in keeping a constant speed. Seventy miles per hour is the speed limit on I-75, 1-10, and I-95 in Florida, Georgia, and South Carolina. Technically, I should average 70 mph each hour while traveling on the interstates. Since road repairs, traffic, and fender benders decrease my speed, it is virtually impossible to maintain 70 mph unless I exceed the speed limit.

With all this fanfare eradicated, let us examine distance-rate-time (DRT) problems under ideal situations.

Example 1: How long will it take someone to drive 350 miles traveling at an average of 50 mph?

Solution:

Distance = Rate • Time

$$350 \text{ miles} = \frac{50 \text{ miles}}{\text{hour}} \bullet x \text{ hours}$$

$$\frac{350}{50} = \frac{50x}{50}$$

$$7 \text{ hours} = x$$

Example 2: I kept track of the miles traveled every hour: 50, 70, 67, 63, 40, and 64. The total distance was 354 miles (50 + 70 + 67 + 63 + 40 + 64). What is the average speed per hour of my trip from Shell Point, FL, to Hilton Head Island, SC, if it took me 6 hours?

Solution:

$$D = R \bullet T$$

$$354 = R \bullet 6 \text{ hours}$$

$$\frac{354}{6} = \frac{6R}{6}$$

$$59 \text{ mph} = R$$

Example 3a: When I traveled from Naples to Shell Point, the total distance was 445 miles. I left Naples at 8 AM, traveled 50 miles the first and last hour and arrived at Shell Point at 4 PM. What was my average mph between 9 AM and 3 PM?

Solution:

$$D = R \bullet T$$

$$445 \text{ miles} - 100 \text{ miles} = R \bullet 6 \text{ hours}$$

$$\frac{345}{6} = \frac{6R}{6}$$

$$57.5 \text{ mph} = R$$

Example 3b: If I used 15 gallons of unleaded gas to travel the 445 miles, what kind of miles per gallon did I average?

Solution:

$$\frac{\text{Distance}}{\text{gallons}} = \text{mi/gal}$$

$$\frac{445}{15} = 29\frac{2}{3}\ \text{mi/gal}$$

Example 3c: If my 15 gallons of gas cost $29.67, what was my cost per mile?

Solution:

$$\frac{\$29.67}{445} = \$0.06\frac{2}{3}\ \text{per mile}$$

DAY 13

REVISITING FACTORING

ON DAY 8, PRIME NUMBERS WAS the focus. During the discussion, I alluded that sooner or later **factoring** would get my attention. So here it goes. It will encompass at least several days. Onward and upward!

You will recall that there are five pairs of factors for 48. They are: 1 and 48, 2 and 24, 3 and 16, 4 and 12, 6 and 8. How can factoring fundamentals help in math? It helps when trying to "factor" a binomial (2 terms), a trinomial (3 terms), and a polynomial (more than 3 terms).

Let us begin by factoring a binomial. That is two terms. Examples of a term: x, 3x, $4x^2$.

Example: $2x + 6b$

Solution:

2 is a factor of itself and also 6.

$$\frac{2x}{2}+\frac{6b}{2}=x+3b$$

When we take a 2 out of $2x + 6b$, we end up with $2(x + 3b)$.

This mathematics process of factoring is the removal of a common monomial (one term) factor. From the above example, I essentially took out a 2 from 2x, leaving an x, and took out a 2 from 6b, leaving 3b. The final result: $2x + 6b = 2(x + 3b)$.

More binomial factoring problems to practice:

a. $16x^2 + 64y^2$
b. $4a^3b^2c^4 - 18a^2b^3c^3$
c. $9x^2 - 25y^2$
d. $8 + 24a^2$
e. $12x^4y^2 - 32a^2x^4$

Several trinomials with common monomial factor:

f. $2x^2 + 6x + 10$
g. $3a^2x + 6ax - 12x$

Let's examine each situation.

a. What common factor is present in $16x^2 + 64y^2$? Certainly not the x^2 nor y^2. Okay, let's look at 16 and 64. This is where factoring knowledge comes in handy.

   Factors of 16 are: 1 and **16**, 2 and 8, 4 and 4

   Factors of 64 are: 1 and 64, 2 and 32, 4 and **16**, 8 and 8

   The largest common factor of both 16 and 64 is 16. Therefore, 16 is the common monomial factor (CMF). Removing the 16 from both terms, ***and*** placing it outside the parenthesis, our answer is $16(x^2 + 4y^2)$.

b. What common factor is present in $4a^3b^2c^4 - 18a^2b^3c^3$? At least a 2, a, b, and c. How many a's, b's, and c's? Pick a minimum of each term. For a, that would be a^2; for b, it would be b^2; and for c, it's c^3. Therefore, the CMF would be $2a^2b^2c^3$.

Removing the CMF from the original problem, what remains? In the first term it is 2ac and in the second terms, 9b.

The final answer is $2a^2b^2c^3(2ac - 9b)$.

c. There is no CMF in $9x^2 - 25y^2$. However, this binomial is an example of a perfect square term minus another perfect square term. One needs to take the square root of each term, 3x and 5y, and have opposite signs for each factor. The final answer is $(3x + 5y)(3x - 5y)$. More about square minus square, cube plus cube, and cube minus cube on Day 14.

d. $8 + 24a^2$ has a CMF in each term. The magic CMF is 8. When 8 is removed from both terms, we end up with 1 and $3a^2$. Yes, we need a 1 to replace the 8. So the correct answer is $8(1 + 3a^2)$.

e. $12x^4y^2 - 32a^2x^4$ has a CMF of $4x^4$. Removing the $4x^4$ from both terms leaves $3y^2 - 8a^2$. The final answer is $4x^4(3y^2 - 8a^2)$.

f. In $2x^2 + 6x + 10$, 2 is the CMF in all three terms. Completely factored, $2(x^2 + 3x + 5)$ is the final answer.

g. Factoring a 3x from the original problem, $3a^2x + 6ax - 12x$, we get $3x(a^2 + 2a - 6)$ which is fully factored.

DAY 14

FACTORING A BINOMIAL

That is either a square minus a square, a cube plus a cube, or a cube minus a cube.

FANTASTIC TITLE FOR DAY 14! CAN we fathom all those binomial squares and cubes? First, one needs to know perfect square numerals. Here's some help. 1,4, 9, 16, 25, 36, 49, 64, 81, and 100 are the first ten numbers squared. These squares are numbers to know. Then the cubes. 1, 8, 27, 64, 125, 216, 343, 512, 729, and 1000 are the first ten cubes. Wow! When you have *memorized* the above facts, we can proceed on our road trip. Don't be too long, cause practice makes perfect.

Here are some practice problems with answers:

Example 1: $27x^3 + 64y^6$ cube plus cube

Solution: $\square + \square$

There is a formula for cubes. Here's the scoop:

$$a^3 + b^3 = (a + b)(a^2 - ab + b^2)$$
$$a^3 - b^3 = (a - b)(a^2 + ab + b^2)$$

So let's try to factor $27x^3 + 64y^6$. The cube root of $27x^3$ is $3x$ and the cube root of $64y^6$ is $4y^2$. So the first factor of the answer is $(3x + 4y^2)$. Now let's figure out the second factor. Using the factor $(3x + 4y^2)$, square the first term $(3x)$, which is $9x^2$. Then, multiply the two terms together, $(3x)(4y^2)$, which is $12xy^2$. Finally, square the second term $(4y^2)$ for $16y^4$. So, putting this altogether, the final answer is

$$(3x + 4y^2)(9x^2 - 12xy^2 + 16y^4).$$

Example 2: $8a^3 - 125b^3$ cube minus cube

Solution: □ − □

Take the cube root of $8a^3$ and $125b^3$, getting $2a$ and $5b$, respectively. The first factor is $(2a - 5b)$. Use this factor to arrive at the second factor: square the first term, multiply the two terms together, and square the second term:

$$(2a)^2 + (2a)(5b) + (5b)^2 =$$
$$4a^2 + 10ab + 25b^2.$$

Putting the two factors together to get:

$$(2a - 5b)(4a^2 + 10ab + 5b^2)$$

which is the correct answer.

Example 3: $100x^4 - 25y^2$ square minus square

Solution: $\square - \square$

Day 13, practice problem c, was a square minus square factor problem. Again we have a square minus square problem. We must first check for a CMF and, low and behold, there is one. The CMF is 25. Removing the CMF, we arrive at $25(4x^4 - y^2)$. The second factor $(4x^4 - y^2)$ is a square minus square and needs to be factored further. We need to take the square root of each term, resulting in:

$$(2x^2 + y)(2x^2 - y)$$

The final answer is $25(2x^2 + y)(2x^2 - y)$.

Remember the formulas!

$$a^2 - b^2 = (a + b)(a - b)$$
$$a^3 + b^3 = (a + b)(a^2 - ab + b^2)$$
$$a^3 - b^3 = (a - b)(a^2 + ab + b^2)$$

Also, L⊙⊙K for a common monomial factor every time you factor.

So far for factoring:

1. Common monomial factor (CMF)
2. If a binomial (2 terms):

   square minus square
   cube plus cube
   cube minus square

What's next on the road trip?

DAY 15

WHAT'S A FRACTION?

AFTER A NASCAR™ ROAD STOP FOR a new fuel pump, I'm back on the highway heading north. Or should I say from the garage back onto the track. Whether it's Talladega, Richmond, or Dover, the ride seems smooth and fuel efficient.

Today, let us discuss the merits of working with fractions. Some math instructors insist that this arena is the student's greatest weakness. While others claim that fractions can be understood with a circle or, better still, a pie, most individuals agree that $\frac{a}{b}$, or a rational number, is unnecessary in today's thriving economy. I totally disagree with the latter.

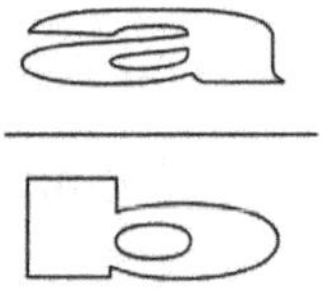

As culinary arts majors, civil engineering students, or quantitative chemistry enthusiasts proclaim, understand the meaning of a fraction can truly make their life easier.

Try to halve a cooking recipe or draw a scale to represent a distance where inches equal feet (or yards, or miles) and one can easily see that understanding fractions in everyday life is essential.

Why not include the idea of a proportion? A proportion is comparing two fractions that equal each other. I will save ratio and proportion for another day and concentrate of fractions for now.

There is the concept of reducing which tries to get the numbers into PRIMES. Yes, there's that prime word again. Knowing the first few primes, like 2, 3, 5, 7, 11, 13, 17, and 19, makes your life easier.

When you reduce a fraction, try to remember to factor out primes, beginning with 2 and working up the line. Odd or even to start. Not too hard. Then add up the digits, and if the sum is divisible by 3, then 3 is a factor. Hey, we're using the rules of divisibility! If it ends in 5 or 0, then 5 is a factor (prime, too) which can be removed. I suggest you work 2, 3, and 5 until you get tired.

If and when the fraction(s) is (are) reduced, it is a lot easier to add and subtract, which requires getting a common denominator. The lower the common denomination when adding or subtracting fractions saves lots of reducing to get a simplified answer.

Example 1: $\frac{3}{8}+\frac{4}{10}+\frac{1}{5}$

Solution:

Reduce $\frac{4}{10}$ to $\frac{2}{5}$ then it's $\frac{3}{8}+\frac{2}{5}+\frac{1}{5}$. Add the two fractions with the same denominator, $\frac{2}{5}+\frac{1}{5}=\frac{3}{5}$. Now add $\frac{3}{8}+\frac{3}{5}$. We need to find the least common denominator, which is 40. Convert $\frac{3}{8}$ to $\frac{?}{40}$. Divide 8 into 40 and multiply the result by 3, which gets you 15, or $\frac{15}{40}$. Now convert $\frac{3}{5}$ to $\frac{?}{40}$. Divide 5 into 40 and multiply the result by 3, which gives 24, or $\frac{24}{40}$. You now have two fractions with the same denominator. Keep the denominator and add the tops (numerators): $\frac{15}{40}+\frac{24}{40}=\frac{39}{40}$.

If you are working with mixed numbers, ***add*** the whole numbers, then work the fractions separately. However, in subtraction, keep the whole numbers around since you may need to borrow.

I have two examples to show how you can win, win, win. Or, as we say, "cheat".

Example 2: $14\frac{3}{8}-6\frac{5}{8}$

Solution:

We need to borrow from the 14, just 1 will do so the whole number becomes 13. Now we have $13\frac{\square}{8}$. What to put in the box? Add the numerator and denominator of the original fraction to get 11 and place that in the box: $13\frac{11}{8}$. Now we can subtract: $13\frac{11}{8}-6\frac{5}{8}=7\frac{6}{8}$. One more thing. $\frac{6}{8}$ can be reduced since both numbers are even: $\frac{6}{8}=\frac{3}{4}$. Therefore the correct answer is: $7\frac{3}{4}$.

Example 3: $19-8\frac{2}{5}$

Solution:

One can't subtract $\frac{2}{5}$ from nothing so borrow 1 from the 19 which makes it 18. Now the problem is $18\frac{\square}{\square}$?. What goes in the box? In this problem we are working with 5[ths], so $1 = \frac{5}{5}$. Now we're ready to roll: $\frac{5}{5} - \frac{2}{5} = \frac{3}{5}$ and 18 – 8 = 10. And the answer is: $10\frac{3}{5}$!

VIOLA! SMILE!

Remember these tidbits for addition and subtraction of fractions.

Try the following adding and subtracting problems to see if it is going great.

1. $\frac{4}{5} + \frac{2}{3}$
2. $3\frac{3}{4} + 1\frac{7}{10}$
3. $5\frac{2}{7} - 2\frac{9}{14}$
4. $2\frac{5}{6} + 1\frac{7}{10} + 3\frac{2}{5}$
5. $4\frac{1}{2} - 3\frac{2}{3}$
6. $7\frac{5}{12} - 3\frac{2}{3}$

DAY 16

PART, WHOLE, AND PERCENT

ON DAY 15, I MENTIONED THE idea of a ratio and/or proportion. Let us examine how to find a part, the whole, or what percent of something is that.

(?? %)
A is what percent of **B**

A is the part.
B is the whole.
?? refers to the percent.

Some examples follow.

Example 1: 16 is 40% of what?

Solution:

The setup goes like this:

$$\boxed{\frac{\text{Part}}{\text{Whole}} = \frac{\%}{100}}$$

Convert 40% to 40: $\frac{16}{x} = \frac{40}{100}$ and cross multiply.

$16 \cdot 100 = 40 \cdot x$ divide by 40

$$\frac{16 \cdot 100}{40} = \frac{40 \cdot x}{40}$$

$40 = x$

Cheat?? Part • 100 = Whole • Percent

$16 \cdot 100 = w \cdot 40$

$40 = w$

Example 2: What is 70% of 200?

Solution:

$$\frac{P}{W} = \frac{\%}{100}$$

$$\frac{P}{200} = \frac{70}{100}$$

P=140

or

$P \cdot 100 = W \cdot \%$

$P \cdot 100 = 200 \cdot 70$

$100P = 14000$

$1ØØP = 140ØØ$

$P = 140$

Remember, when dividing – you can cancel zeros on both sides of the equal sign, one-for-one!

Example 3: 24 is what percent of 80?

Solution:

$$\frac{P}{W}=\frac{\%}{100}$$
$$\frac{24}{80}=\frac{\%}{100}$$
$$30=\%$$

or

$$P \cdot 100 = W \cdot \%$$
$$24 \cdot 100 = 80 \cdot \%$$
$$240\emptyset = 8\emptyset \cdot \%$$
$$30 = \%$$

Example 4: 60% of 20 is what?

Solution: Remember that the number after "of" is the whole, so we are to find the part.

$$\frac{P}{W}=\frac{\%}{100}$$
$$\frac{P}{20}=\frac{60}{100}$$
$$P \cdot 100 = 20 \cdot 60$$
$$1\emptyset\emptyset P = 12\emptyset\emptyset$$
$$P = 12$$

Try some now!

1. What is 5% of $22.60?
2. 110% of 450 is what number?
3. What percent of 300 is 3.6?
4. 50 is 20% of what number?
5. 4.5% of what number is 13.5?
6. 23 is what percent of 92?
7. 8 is 20% of what number?
8. What percent of 150 is 3?

9. 15% of 180 is what number?
10. 10 is what percent of 80?
11. 21 is 70% of what number?
12. What percent of 63 is 21?
13. What percent of 36 is 45?
14. 4.2 is 75% of what number?
15. 78 is 15% of what number?

DAY 17

MORE FACTORING

DAYS 13 AND 14 CONCENTRATED ON factoring. First, we looked for a common monomial factor (CMF) in a group of terms. Then we selected binomials (two terms) for investigation. Basically, there is (i) square minus square, (ii) cube plus cube, and (iii) cube minus cube.

Now we will look at polynomials (many terms). The first group of polynomials is four terms. When we see four terms, we should think "grouping".

So to summarize to this point, we follow these guidelines:

1. CMF – common monomial factor
2. 2 terms – binomials (square minus square, cube plus cube, cube minus cube)
3. 4 terms – grouping

Here is an example for factoring by grouping and how it is factored:

$8a^2 - 10ab + 12ab - 15b^2$

Solution:

Step 1: Group the first two terms and the last two terms.

$(8a^2 - 10ab) + (12ab - 15b^2)$

Step 2: Find a CMF in $8a^2 - 10ab$. The CMF is 2a. Now remove, by dividing, the 2a from both terms, which leaves: $2a(4a - 5b)$. With this idea in mind, find a CMF in $12ab - 15b^2$. It is 3b. When dividing 3b into $12ab - 15b^2$, the answer is $4a - 5b$. So factored, it is $3b(4a - 5b)$.

Let's review what we have so far:

$8a^2 - 10ab + 12ab - 15b^2 =$
$(8a^2 - 10ab) + (12ab - 15b^2) =$
$2a(4a - 5b) + 3b(4a - 5b)$.

Now there is a CMF in both factored terms. It is $4a - 5b$. Extract $4a - 5b$ from both terms. What remains is $2a + 3b$.

Completely factored,

$8a^2 - 10ab + 12ab - 15b^2 =$

$(4a - 5b)(2a + 3b)$.

Using the FOIL method, let's check our work.

<u>F</u>irst <u>O</u>uter <u>I</u>nner <u>L</u>ast

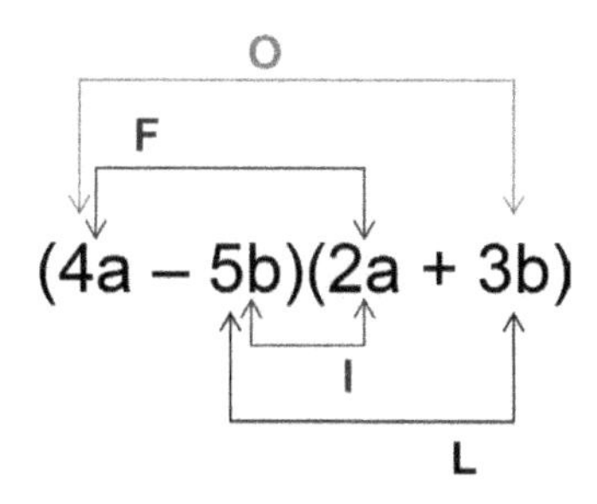

First terms:	$4a \bullet 2a = 8a^2$
Outer terms:	$4a \bullet 3b = 12ab$
Inner terms:	$-5b \bullet 2a = -10ab$
Last terms:	$-5b \bullet 3b = -15b^2$

$$8a^2 + 12ab - 10ab - 15b^2$$

Let's review several problems about grouping.

1. $(a + b)(2 + c) + x(2 + c)$
2. $x^2(y + 2) - 5(y + 2)$
3. $x(x - 2) - 6(x - 2)$
4. $pq + 5q + 2p + 10$
5. $2xy + 3y + 2x + 3$
6. $2a^2 - 4a + 3ab - 6b$
7. $x^3 + 3x^2 - 5x - 15$
8. $6ax + 24x + a + 4$
9. $2x^2 - 10x + 3xy - 15y$
10. $2p + ap + 2q + aq$
11. $2x^3 + x^2 + 8x + 4$
12. $2a^3 - a^2 - 10a + 5$
13. $4x^2 - 2xy - 7yz + 14xz$
14. $7t(n - 4) - (4 - n)$
15. $x^2 - 16y^2 + 2x + 8y$

DAY 18

WORK PROBLEMS

As we roll along, we find more story problems in every nook and cranny. We need to stop at this nook and examine "work" problems.

These types of problems involve working with fractions. On Day 15, we studied fractions. Adding and subtracting was on our agenda that day. Now we utilize our expertise with fractions in solving work problems.

Let us examine several work (story) problems that are solved using fractions.

Example 1: Jack can paint the exterior of a two-story colonial house in 7 days. If Tom paints that same house in 10 days by himself, how long will it take both Jack and Tom, working together, to paint the exterior of the two-story colonial house?

Solution:

Let x = working together.

$$\frac{x}{7} + \frac{x}{10} = 1$$

Why equal to 1? That means the task was completed.

Let's get rid of the fractions by multiplying everything by the least common denominator. In this problem, the LCD is 70.

$$\frac{70 \bullet x}{7} + \frac{70 \bullet x}{10} = 70 \bullet 1$$
$$10x + 7x = 70$$
$$17x = 70$$
$$x = \frac{70}{17} = 4\frac{2}{17} \text{ days}$$

Example 2: You are filling a swimming pool with water from a hose. It takes 36 hours to completely fill that swimming pool. However, last fall, when you drained and winterized the pool, you forgot and left the drain open. It took 48 hours to drain the pool in the fall. Our problem is to find out how long it takes to fill the pool with ***both*** the drain open and the hose filling the pool.

Solution:

Let x = working together.

$$\frac{x}{36} - \frac{x}{48} = 1$$

This time the drain removes the water, so we need to subtract it. The LCD for this problem is 144.

$$\frac{144 \bullet x}{36} - \frac{144 \bullet x}{48} = 144 \bullet 1$$

$$4x - 3x = 144$$
$$x = 144 \text{ hours or 6 days!}$$

JUST A THOUGHT! ANOTHER WAY TO SOLVE A WORK PROBLEM WITH A HEADS UP PLAY! IT IS "CHEATING"!

Now try it using this formula:

$$\frac{\text{Product of the two individuals or objects}}{\text{Sum (Difference)}}$$

Example1: $\dfrac{\text{Jack} \bullet \text{Tom}}{\text{Jack} + \text{Tom}} = \dfrac{7 \bullet 10}{7+10} = \dfrac{70}{17} = 4\dfrac{2}{17}$ days

Example 2: $\dfrac{\text{Hose} \bullet \text{Drain}}{\text{Jack} - \text{Hose}} = \dfrac{36 \bullet 48}{48-36} = = 144$ hours

Example 3: Ralph and Isaac together mow a lawn in 5 hours. When Ralph mows it alone, it takes him 8 hours. How long will it take Isaac, working alone, to mow the lawn?

Solution:

Let y = Isaac working alone

$$\frac{\text{working together}}{\text{Ralph}} + \frac{\text{working together}}{\text{Isaac}} = 1$$

$$\frac{5}{8} + \frac{5}{y} = 1$$

$$\frac{5}{y} = 1 - \frac{5}{8}$$

Subtract $\frac{5}{8}$ from by sides of the equation

$$\frac{5}{y} = \frac{3}{8}$$

Here we cross-multiply.

$$3y = 40$$

$$y = \frac{40}{3} = 13\frac{1}{3} \text{ hours}$$

An alternate way of finding the solution would be to find the least common denominator of the fractions.

LCD: 8y

$$\frac{8y \bullet 5}{8} + \frac{8y \bullet 5}{8} = 8y \bullet 1$$

Multiply ***ALL*** terms by the LCD. Don't forget the right side of the equation which has **1** as its denominator.

$$5y + 40 = 8y$$

Subtract "5y" from both sides of the equation.

$$40 = 3y$$

$$y = \frac{40}{3} = 13\frac{1}{3} \text{ hours}$$

Problems:

1. It takes Bill 6 hours to paint his living room. If it takes Ann 10 hours to paint the same living room, how long

does it take both of them to paint the living room when working together?

2. When filling a swimming pool, it takes 12 hours. To drain the same pool, it takes 16 hours. If both pipes are open, how long does it take to fill the pool?
3. Al can do a job in 30 minutes. Bill can do the same job in 40 minutes and Casey can do the exact same job in 1 hour. If they work together, how long will it take them?
4. Spartan Cleaning Service can clean a school in 8 hours. If the janitors take twice as long to clean the school, how long will it take to complete the job when the Spartan Cleaning Service and janitors are working together?

DAY 19

RATIO AND PROPORTION

RAINDROPS KEEP FALLING ON MY AUTOMOBILE as I roll back to Hilton Head Island. Today is not the day to meander on the golf course. Too many lights for daylight hours! Could that be lightning?

$$\frac{a}{b} = \frac{c}{d}$$

Indoors today. Let us discuss ratios and proportions today. They go hand-in-hand. Two ratios make a proportion.

Let us examine how ratios are created and then we will venture into proportions.

Pretend you are vacationing at the beach. Of the seven days at the beach, it rains two days. That gives you a ratio of 2 to 7, 2 days of rain to 7 days at the beach.

The ratio is written like a fraction: $\frac{2}{7}$. If we use that ratio to give us a sense of how many days could it rain in two weeks we could set up a proportion: $\frac{2}{7} = \frac{x}{14}$, where 2 weeks = 14 days.

Remember on Day 5, we set up a proportion and solved it by cross-multiplication.

$$(2)(14) = (7)(x)$$
$$28 = 7x$$
$$4 = x$$

This is a 50-50 chance that it could rain 4 days out of 14 days while you are at the beach. As we all know, the weather is sometimes unpredictable.

Gas mileage is an excellent application of ratios. If someone drives 300 miles in 6 hours, their average speed is $\frac{300 \text{ miles}}{6 \text{ hours}}$ or 50 mph.

Also, for traveling the 300 miles, we needed 12 gallons of gasoline. How many miles per gallon is that?

$$\frac{300 \text{ miles}}{12 \text{ gallons}} = 25 \text{ miles/gallon}$$

Here are some representative proportion problems to practice.

1. If it takes 5 hours to travels 260 miles, how long will it take to travel 468 miles?
2. If a tree 20 feet tall costs a shadow 5 feet long, how tall is a building that casts a 15-foot shadow simultaneously?
3. If $\frac{1}{8}$ inch on a map represents 4 miles, 2 inches represents how many miles?
4. The school referendum passed by 7 to 5 majority. If 2135 people voted for the referendum, how many voted against it?
5. If four CDs costs $22, how many CDs can I buy with $50?

DAY 20

REVIEW OF FACTORING

THE WEATHER IS ABSOLUTELY AMAZING. DAY 19 saw raindrops in the morning and sunshine by early afternoon. Today the bright sunshine and 55° temperature makes one think about hitting the links. Prior to that outing, we need to review factoring with these concepts in mind:

#1: Look for a CMF.

#2: If it's a binomial, check for square minus square, cube plus cube, or cube minus cube.

#3: Grouping with a four-term polynomial.

Now onto the practice to see if we can move forward.

Example 1: $x^2(y + 2) - 9(y + 2)$

Solution:

Find a CMF if present. The CMF is $(y + 2)$. Remove the CMF from both terms, arriving at $(y + 2)(x^2 - 9)$. ARE WE DONE?

NO! Why not? Correct, you can STILL factor the $(y^2 - 9)$. It is a square minus a square. $(x^2 - 9)$ factors into $(x + 3)$ and $(x - 3)$.

Result: $(y + 2)(x + 3)(x - 3)$

TRICKS OF BEING A MATH INSTRUCTOR!

I often slide that problem into a factoring exam and you guessed it! The cat wins! That's me!! Purr! Playing Sudoku, looking out my picture window at the 3rd tee and 6th green at Harbourtown Golf Club, and writing about math frustrations, life is fun!

Example 2: $x^4 + x^3 + 8a^3x + 8a^3$

Solution:

Group the first two terms and the last two terms:

$$(x^4 + x^3) + (8a^3x + 8a^3)$$

L⊙⊙K for a CMF in both groups and remove it.

$$(x^4 + x^3) = x^3(x + 1)$$
$$(8a^3x + 8a^3) = 8a^3(x + 1)$$

We now have: $x^3(x + 1) + 8a^3(x + 1)$.

Again, look for a CMF in both groups and remove it, $(x + 1)$ is common to both. Now we have:

$$(x + 1)(x^3 + 8a^3)$$

But $(x^3 + 8a^3)$ is a cube plus cube. It factors into $(x + 2a)(x^2 - 2ax + 4a^2)$.

Final answer: $(x + 1)(x + 2a)(x^2 - 2ax + 4a^2)$.

Now we are handling the factoring dilemmas.

Example 3: $x^4 - 16x^2$

Solution:

We have a CMF: x^2. Removing x^2 from both terms we get:

$x^2(x^2 - 16)$. Is this fully-factored? No.

$(x^2 - 16)$ is a square minus a square. It factors into $(x + 4)$ and $(x - 4)$.

Final answer: $x^2(x + 4)(x - 4)$

Example 4: $x^3(x^2 - 9) - y^3(x^2 - 9)$

Solution: Double trouble!!

First factor out the CMF of $(x^2 - 9)$. What remains is $(x^3 - y^3)$. Giving us $(x^2 - 9)(x^3 - y^3)$. Both of which need factoring again. $(x^2 - 9)$ is a square minus square. $(x^3 - y^3)$ is a cube minus cube.

$(x^2 - 9) = (x + 3)(x - 3)$

$(x^3 - y^3) = (x - y)(x^2 + xy + y^2)$

Putting it altogether, we have:

$(x + 3)(x - 3)(x - y)(x^2 + xy + y^2)$

Example 5: $16x^4 - 256y^4$

Solution:

Remove the CMF: $16(x^4 - 16y^4)$.

Factor $(x^4 - 16y^4)$, a square minus square:

$(x^2 + 4y^2)(x^2 - 4y^2)$.

Factor $(x^2 - 4y^2)$, another square minus square: $(x + 2y)(x - 2y)$. Finally we have our answer:

$16(x^2 + 4y^2)(x + 2y)(x - 2y)$.

That was lots of work!

One more for good measure!

Example 6: $a^4x^4 - 81b^4y^4$

Solution:

There is no CMF. But it is a square minus square.

$a^4x^4 - 81b^4y^4 = (a^2x^2 + 9b^2y^2)(a^2x^2 - 9b^2y^2)$

However $(a^2x^2 - 9b^2y^2)$ is another square minus square.

$(a^2x^2 - 9b^2y^2) = (ax + 3by)(ax - 3by)$

Fully factored we have:

$(a^2x^2 + 9b^2y^2)(ax + 3by)(ax - 3by)$

That's enough for now!

5 Factoring?

1. $12y^3 + 36y^2 + 27y$
2. $25a^2 - 4b^2 - 4b - 1$
3. $(e + 3)^2 - 7(e + 3) + 10$
4. $27z^3 + 3z + 18z^2$
5. $6c^2 + c - 15$

DAY 21

SOLVING DECIMAL EQUATIONS

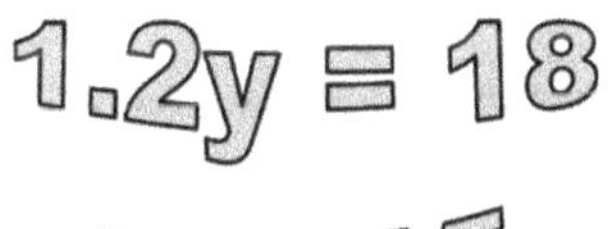

ANOTHER DAY. OVERCAST WITH A SLIGHT chance of showers. First let's do some algebra prior to hitting the links.

Today's topic is working with decimals in solving linear equation. We will work our way up the ladder in difficulty. So here it goes!

Example 1: $0.6x = 15$

Solution:

Clear out the decimal point by multiplying both sides of the equal sign by 10.

$0.6x(10) = 15(10)$

$6x = 150$ Divide both sides by 6.

$x = 25$

Check it out: $0.6(25) = 15$ ✓ AOK!

Example 2: $1.4x + 6 = -22$

Solution:

Again, multiply all three terms by 10.

$1.4x(10) + 6(10) = (-22)(10)$

$14x + 60 = -220$ Subtract 60 from both sides.

$14x = -280$ Divide by sides by 14.

$x = -20$

Check it out:

$1.4(-20) + 6 = -22$

$-28 + 6 = -22$

$-22 = -22$ ✓ AOK!

Note: Students sometime like to get variables and constants on opposite sides of the equation. It is your choice. See Example 3.

Example 3: $2.4x + 0.6x = 50.46 + 12.3$

Solution:

I suggest adding 2.4x and 0.6x on the left side. Also, adding 50.46 and 12.3 of the right side.

$3x = 62.76$ Divide both sides by 3

$x = 20.92$

Check it out:

$2.4(20.92) + 0.6(20.92) = 50.46 + 12.3$

$50.208 + 12.552 = 62.76$

$62.76 = 62.76$ AOK!

Three for three! Keep on rolling!

Example 4: $3.02x + 15.4 = 1.6x + 24.275$

Solution: (use your calculator)

Multiply **all** four terms by 1000 to remove all the decimals. Why 1000? Because 24.326 needs to move three places to the right. That would be 1000.

$$3.020\,x + 15.400 = 1.600\,x + 24.275$$

Subtract 15400 from both sides.

$$\begin{array}{r} 3020x + 15400 = 1600x + 24275 \\ -15400 \qquad\qquad -15400 \\ \hline \end{array}$$

Subtract 1600x from both sides.

$$\begin{array}{r} 3020x = 1600x + 8875 \\ -1600x \quad -1600x \qquad\quad \\ \hline \end{array}$$

$$1420x = 8875$$

Divide both sides by 1420.

$$\frac{1420x}{1420} = \frac{8875}{1420}$$

$$x = 6.25$$

Check it out:

$3.02(6.25) + 15.4 = 1.6(6.25) + 24.275$

$18.875 + 15.4 = 10 + 24.275$

$34.275 = 34.275$ AOK!

Example 5: $0.14x - 3.81 = 0.07x + 1.23$

Solution:

Multiply all four terms by 100.

$14x - 381 = 7x + 123$

Subtract 7x from both sides and add 381 to both sides.

$$\begin{array}{rcl} 14x - 381 & = & 7x + 123 \\ -7x + 381 & & -7x + 381 \\ \hline \end{array}$$

$7x = 504$ Divide both sides by 7.
$x = 72$

Check it out!

In this example, we have added and subtracted simultaneously.

Try these for experience.

1. $2.1x - 3.2 = -8.4x - 45.2$
2. $6.4 - 4.2x = 16.8x + 90.4$
3. $2.418x = -12.09$
4. $34.3x - 5.95x = 283.5$
5. $2.18x + 6x - 4.3225 = 8.645 - 0.465x$
6. $30.7x - 18x = 76.2$
7. $39.4 = 7.32 + 3.208x$
8. $2.48x + 8x = 92.28 - 4.9x + 30.76$

DAY 22

FACTORING A TRINOMIAL WITH AN x^2 COEFFICIENT OF 1

(Trial and Error)

ON DAYS 13, 14, 17, AND 20, we discussed factoring. On Day 20, it was a review. If you have taken some time off between Day 20 and today, go back and review it.

Factoring a trinomial whose coefficient of the squared variable is +1, we can use the trial and error theory.

Example 1: $x^2 + 4x + 3$

Solution:

Step 1: $(x + \quad)(x + \quad)$ set up this first

Step 2: $(x + 3)(x + 1)$ only factors of 3 are 1 and 3

Step 3: Check it using FOIL!

$F \rightarrow x^2 \qquad \boxed{O \rightarrow x + I \rightarrow 3x} \qquad L \rightarrow 3$

$x^2 + 4x + 3$ GOOD!

Example 2: $x^2 - 4x + 3$

Solution: Notice that this is the same as example 1 except for the signs.

Step 1: $(x - \quad)(x - \quad)$

Step 2: $(x - 3)(x - 1)$

Step 3: Remember to check by using FOIL!

Now we have two locks!

#1: $x^2 + ax + b$, where both signs are positive, must always be $(x + \quad)(x + \quad)$.

#2: $x^2 - ax + b$, where first sign is negative and second sign is positive, must always be $(x - \quad)(x - \quad)$.

However, if we have a two negative-sign trinomial or a positive-sign followed by a negative-sign trinomial, the signs of the factors will be one positive sign and one negative sign. The sign of the middle term will determine the largest factors' sign.

Check out the following examples.

Example 3: $x^2 - 5x - 24$

Solution: This will factor into $(x - 8)(x + 3)$.

Let me explain. The last term is 24 which has several factors: (24)(1), (12)(2), (8)(3), and (6)(4).

(24)(1): gives 25 when added
gives 23 when subtracted

(12)(2): gives 14 when added
gives 10 when subtracted

(8)(3): gives 11 when added
gives 5 when subtracted

(6)(4): gives 10 when added
gives 2 when subtracted

In the original trinomial, the middle term is $-5x$. Therefore we need the factors 8 and 3, where 8 gets the negative sign and 3 gets the positive sign. Therefore, $x^2 - 5x - 24$ factored is $(x - 8)(x + 3)$. Check using FOIL.

Example 4: $x^2 + 5x - 24$

Solution: Same as above example but the signs are reversed. Since we now have a $+5x$ as the middle term the factors are $(x + 8)(x - 3)$. Use FOIL to check!

Now try these practice problems. Do some today, then more tomorrow. Mark this page so you can refer to it easily.

1. $a^2 - 2a - 15$
2. $x^2 + 5x - 36$
3. $b^2 + 4b - 21$
4. $x^2 - 2x - 35$
5. $x^2 + 11x - 26$
6. $c^2 + 3c - 40$
7. $r^2 + 4r - 96$
8. $m^2 + 3m - 54$
9. $y^2 + 3y - 28$
10. $e^2 - 2e - 99$
11. $f^2 + 11f - 80$
12. $a^2 - 9a - 36$
13. $x^2 - 2x - 48$
14. $x^2 - 5x - 50$
15. $x^2 - 17x - 60$

DAY 23

DISTRIBUTIVE PROPERTY

MANY STUDENTS SEE THE DISTRIBUTIVE PROPERTY as a factor exercise. In many ways, that assumption is true.

The Distributive Law of Multiplication Over Addition goes like this:

$$a(b + c) = ab + ac$$

Multiply "a" times both "b" and "c".

Sometimes you can add the numbers within the parentheses if they have the same base.

Example 1: $4(2 + 3) = 4(5) = 20$

or $4(2 + 3) = 4(2) + 4(3) = 8 + 12 = 20$

However, if the terms in the parentheses are not the same, then

Example 2: $4(2 + x)$ needs to be run out,

$4(2) + 4(x) = 8 + 4x$

It even works for distribution of multiplication over subtraction.

Example 3: $3(8 - 6) = 3(2) = 6$

or $3(8) - 3(6) = 24 - 18 = 6$

Example 4: $3(8x - 6) = 3(8x) - 3(6) = 24x - 18$

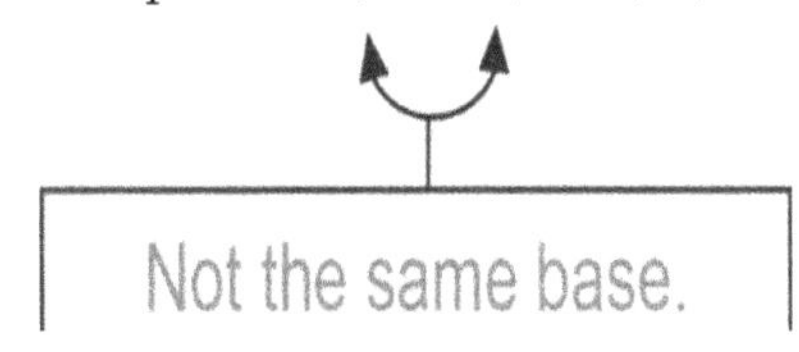

When you reverse the order and have something like $12x - 6$, you can factor out a 6 and the result is $6(2x - 1)$. This could and ***will*** help us in factoring.

A WORD OF CAUTION!!!!
WATCH OUT FOR NEGATIVES IN PARENTHESES!!!

$$a(x - b) = ax - ab$$
$$-a(x - b) = -ax + ab$$
$$a(-x - b) = -ax - ab$$
$$-a(-x + b) = ax - ab$$

Example 5: $6(x - 4)$ gives us $6x \boxed{-} 24$.

Example 6: $4x - 3(x + 5) = 4x \boxed{-} 3x \boxed{-} 15$

or $x - 15$

The –3 effects both the x and the 5.

PLEASE BE CAUTIOUS!!!
ALWAYS CAREFUL!!!
AND CORRECT!!!

Some practice:

	Expand	Reverse it (factor)
a.	$5(x + 3)$	$5x + 15$
b.	$-4(a + 5)$	$-4a - 20$
c.	$3(-4 + x)$	$-12 + 3x$
d.	$2(y - 4)$	$2y - 8$
e.	$6(z - 3)$	$6z - 18$
f.	$-8(4 + y)$	$-32 - 8y$

	Reverse it (factor)	Expand
g.	$6x + 18$	$6(x + 3)$
h.	$3a + 3b$	$3(a + b)$
i.	$5y - 20$	$5(y - 4)$
j.	$-3x - 21$	$-3(x + 7)$
k.	$4x - 12$	$4(x - 3)$
l.	$12 - 6x$	$6(2 - x)$

	Expand	Reverse it
m.	$-5(a + b)$	$-5a - 5b$
n.	$16(x + 2)$	$16x + 32$
o.	$4(3 + 7)$	$4(10) = 40$
p.	$-2(a - a^2)$	$-2a + 2a^2$
q.	$-x(y + x)$	$-xy - x^2$
r.	$3x(2y + 3z)$	$6xy + 9xz$
s.	$10(-3z + x)$	$-30z = 10x$
t.	$7y(x - z)$	$7xy - 7yz$
u.	$-4(8 - 7a)$	$-32 + 28a$
v.	$11(3a - 4b)$	$33a - 44b$
w.	$-12 (e + 4f)$	$-12e - 48f$
x.	$5(x - 5y)$	$5x - 25y$
y.	$-3x(y - 5z)$	$3xy + 15xz$
z.	$-10x(3x + 5y)$	$-30x^2 - 50xy$

DAY 24

SOLVING FRACTIONAL EQUATIONS

ANOTHER ARENA OF DIFFICULTY WITH ALGEBRA is solving equations containing fractions. None with standing, "**FRACTIONS**" is the most dreaded word in mathematics.

Let us review several examples to alleviate those misapprehensions.

Example 1: $\frac{1}{3}x = 2$

Solution:

Multiply both sides of the equation by 3, the reciprocal of $\frac{1}{3}$.

$3 \cdot \frac{1}{3}x = 2 \cdot 3$ The threes on the left cancel.

$x = 6$

Check your solution in the original equation:

$\frac{1}{3} \cdot 6 = 2$ ✓

Example 2: $-\frac{2}{7}y = 10$

Solution:

Multiply both sides by the reciprocal of $-\frac{2}{7}$, which is $-\frac{7}{2}$.

$$\left(-\frac{7}{2}\right)\bullet\left(-\frac{2}{7}\right) =$$

$$\left(-\frac{7}{2}\right)\bullet\left(-\frac{2}{7}\right)y = 10\bullet\left(-\frac{7}{2}\right)$$

$$y = \frac{10(-7)}{2} = -\frac{70}{2} = -35$$

Checking the solution: $-\frac{2}{7}\bullet(-35) = 10$

$$\frac{70}{7} = 10$$

$$10 = 10 \checkmark$$

Example 3: $x + \frac{1}{4} = 3$

Solution:

Subtract $\frac{1}{4}$ from both sides.

$$x + \frac{1}{4} = 3$$
$$-\frac{1}{4} \quad -\frac{1}{4}$$
$$x = 3 - \frac{1}{4} = 2\frac{3}{4}$$

⟺

$$3 = 2\frac{4}{4}$$
$$-\frac{1}{4} = -\frac{1}{4}$$
$$2\frac{3}{4}$$

Check: $2\frac{3}{4} + \frac{1}{4} = 3$

$$3 = 3 \checkmark$$

Example 4: $x - \frac{3}{5} = \frac{1}{10}$

Solution:

$$\begin{array}{r} \frac{1}{10} = \frac{1}{10} \\ +\frac{3}{5} = +\frac{6}{10} \\ \hline \frac{7}{10} \end{array}$$

Add $\frac{3}{5}$ to both sides.

$$\begin{array}{rr} x - \frac{3}{5} = & \frac{1}{10} \\ +\frac{3}{5} & +\frac{3}{5} \\ \hline \end{array}$$

reduce by 5

$$x = \frac{1}{10} + \frac{3}{5} = \frac{(5)(1)+(3)(10)}{(10)(5)} = \frac{5+30}{50} = \frac{35}{50} = \frac{7}{10}$$

Check: $\frac{7}{10} - \frac{3}{5} = \frac{1}{10}$

$$\frac{(7)(5)-(3)(10)}{(10)(5)} = \frac{35-30}{50} = \frac{5}{50} = \frac{1}{10} \checkmark$$

Example 5: $x + \frac{1}{5} = \frac{2}{3}$

Solution:

$$\frac{2}{3} = \frac{10}{15}$$
$$-\frac{1}{5} = -\frac{3}{15}$$
$$\frac{7}{15}$$

Subtract $\frac{1}{5}$ from both sides.

$$\begin{array}{rcr} x + \frac{1}{5} & = & \frac{2}{3} \\ -\frac{1}{5} & & -\frac{1}{5} \\ \hline \end{array}$$

$$x = \frac{2}{3} - \frac{1}{5} = \frac{(2)(5)-(1)(3)}{(3)(5)} = \frac{10-3}{15} = \frac{7}{15}$$

Check it!! $\frac{7}{15} + \frac{1}{5} = \frac{2}{3}$

$$\frac{(7)(5)+(1)(15)}{(15)(5)} = \frac{35+15}{75} = \frac{50}{75} = \frac{2}{3} \checkmark$$

reduce by 25

Now try these to sharpen your math. Hint: change mixed numbers to improper fractions.

1. $\frac{2}{3}x = 14$

2. $\frac{4y}{3} = -2$

3. $-\frac{8x}{5} = -16$

4. $2\frac{1}{3}x = -14$

5. $-\frac{3}{5}x = 6$

6. $\frac{x}{7} = -5$

7. $x + \frac{2}{3} = 3\frac{1}{4}$

8. $x - 1\frac{1}{2} = 5\frac{2}{3}$

9. $6 - \frac{x}{3} = 4$

10. $5\frac{2}{3} - x = 16$

11. $a + \frac{3}{4} = 2$

12. $x - \frac{2}{5} = \frac{1}{10}$

13. $y + \frac{1}{3} = \frac{3}{5}$

14. $-2x = \frac{6}{7}$

15. $-\frac{5}{6} + y = \frac{7}{18}$

16. $x + \frac{2}{3} + \frac{5}{8} = 12$

17. $y - \frac{7}{10} + \frac{3}{5} = 4\frac{2}{5}$

18. $\frac{2x}{3} = \frac{1}{6}$

hint: COMBINE both factions first

19. $\frac{1}{4}+\frac{2}{x}=\frac{3}{4}$

20. $\frac{3}{y}-\frac{1}{4}=\frac{1}{12}$

21. $\frac{x}{x+2}=\frac{5}{y}$

22. $\frac{9}{x-1}=\frac{1}{2}$

23. $\frac{n}{n-2}=\frac{6}{5}$

24. $\frac{6-x}{4-x}=\frac{3}{5}$

25. $\frac{x-4}{x-2}=2$

26. $\frac{1}{2x}+\frac{2}{3x}=\frac{-7}{36}$

27. $\frac{20}{3x}+\frac{5}{2x}=\frac{5}{6}$

28. $\frac{1+y}{3y}=\frac{1}{y}$

29. $x+\frac{x-2}{8}=20$

30. $\frac{5a+2}{3}=\frac{a-1}{2}$

DAY 25

FACTORING A TRINOMIAL

(Coefficient of $x^2 > 1$)

TODAY IS THE FINAL INSTALLMENT ON understanding how to factor. It's factoring a trinomial whose x^2 coefficient is greater than 1. It involves grouping, setting up four terms to factor. In general, it looks like this: $Ax^2 + Bx + C$.

Let me explain a few things before your try. Remember, we are looking for two factors, when multiplied, giving us the required trinomial. So here it goes!

Example 1: $2x^2 - 7x - 15$

Solution:

The key here is to multiply the x^2 coefficient and the constant. In the above trinomial, that would be the 2 and -15. The result is -30.

Next we examine what two factors of -30 has a sum or difference of -7. (That's the coefficient of the middle term.) Our choices are: $(1)(-30)$, $(2)(-15)$, $(3)(-10)$, $(5)(-6)$, $(6)(-5)$, $(10)(-3)$, $(15)(-2)$, or $(30)(-1)$. From these eight choices, there is one obvious answer: (3)

(−10). Having selected these two factors, we rewrite the trinomial and make it a four-term polynomial: (a) $2x^2 + 3x - 10x - 15$ or (b) $2x^2 - 10x + 3x - 15$. Either way, our answer will be the same.

Let us try (b) first. Group the first two terms and the last two terms:

$$(2x^2 - 10x) + (3x - 15)$$

Next, take a common factor out of both groups: 2x in the first group and 3 in the second group.

Removing the 2x from the two terms in the first group we have: $2x(x - 5)$. Likewise, the 3 is removed from the two terms in the second group, resulting in $3(x - 5)$.

Combining everything we have done so far, we have:

$$2x^2 - 10x + 3x - 15$$

$$(2x^2 - 10x) + (3x - 15)$$

$$2x(x - 5) + 3(x - 5)$$

From the last step, we show a CMF (common monomial factor) in both groups. It is $(x - 5)$. Remove the $(x - 5)$ from both groups. What remains is $(2x + 3)$.

The final answer is $(x - 5)(2x + 3)$.

Check your work by multiplying $(x - 5)$ times $(2x + 3)$.

F	$(x)(2x) = 2x^2$
O	$(x)(+3) = +3x$
I	$(-5)(2x) = -10x$
L	$(-5)(-3) = -15$

Combine O ($+3x$) and I ($-10x$) to get $-7x$.

Final check: $2x^2 - 7x - 15$.

When we try (a) $2x^2 + 3x - 10x - 15$, we should get the same answer.

Separate the $(2x^2 + 3x)$ from the $(-10x - 15)$.

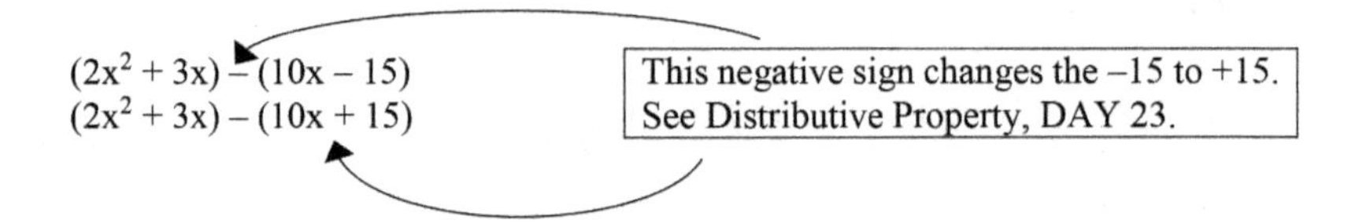

Find a CMF in both groups and remove it. In $(2x^2 + 3x)$, it is "x". In $(10x + 15)$ it is "5". This results in:

$$x(2x + 3) - 5(2x + 3)$$

Remove the CMF from both groups, which is $(2x + 3)$: $(2x + 3)(x - 5)$

NOTE: your can reverse the factors also: $(x - 5)(2x + 3)$.

Therefore, the answer can be $(x - 5)(2x + 3)$ **or** $(2x + 3)(x - 5)$.

Let's try another,

Example 2: $6x^2 + x - 35$.

Solution:

Product $(6 \cdot -35)$ is -210.

Difference is 1.

Factors of -210 are: $(1)(-210)$, $(-1)(210)$, $(2)(-105)$, $(-2)(105)$, $(3)(-70)$, $(-3)(70)$, $(4)(-52)$, $(-4)(52)$, $(6)(-35)$, $(-6)(35)$, $(7)(-30)$, $(-7)(30)$, $(10)(-21)$, $(-10)(21)$, $(14)(-15)$, $(-14)(15)$. Our last try gets us a difference between factors of $+1$.

Rewrite the trinomial as a four-term polynomial: $6x^2 - 14x + 15x - 35$

Group. (Note, put the negative as the second term and the positive as the third terms for grouping the four-term polynomial.)

$(6x^2 - 14x) + (15x - 35)$

Find the CMF for each and remove it:

$2x(3x - 7) + 5(3x - 7)$

Again, find the CMF and remove it.

$(3x - 7)(2x + 5)$

Check the final answer.

F $(3x)(2x) = 6x^2$

O $(3x)(5) = 15x$

I $(-7)(2x) = -14x$

L (–&)(5) = –35

Combining O (15x) and I (–14x) for the middle term of +x.

$6x^2 + x - 35$ voila!

Try this one: $2b^2 + b - 28$
answer: $(2b - 7)(b + 4)$

another: $3x^2 + 25x + 8$
answer: $(3x + 1)(x + 8)$

another: $5x^2 - 31x - 28$
answer: $(5x + 4)(x - 7)$

Take that last one:

Example 3: $5x^2 - 31x - 28$.

Solution:

We know for sure that the factors need to look something like this $(5x \quad)(x \quad)$ because the only factors of 5 are itself and one.

Now examine -28. There are several factors besides itself and one. They are (14)(2) and (7)(4).

Taking the factors (14)(2), we multiply 14 by 5 gives 70, that will be 70 ± 2 or 72 and 68. Whereas the factors (7)(4) would be (7 times 5) $\pm$ 4, or 39 and 31. There is our 31.

Now substitute in 7 and 4 with their signs so $(5x + 7)(x - 4)$ gives $-13x$ for the middle term. Oops. Let's reverse the order: $(5x + 4)(x - 7)$ gives $-31x$ for the middle term. Therefore, $(5x + 4)(x - 7)$ is what we want.

Try the above method (called T & E) for these:

1.	$7x^2 + 23x + 6$	2.	$2x^2 - 11x + 14$
3.	$4x^2 + 17x - 15$	4.	$5x^2 + 13x - 6$
5.	$3x^2 + 13x + 4$	6.	$5x^2 + 6x - 8$
7.	$4y^2 - 11y + 6$	8.	$9x^2 - 21x + 10$
9.	$6x^2 + 5x - 4$	10.	$6n^2 - 11n - 10$
11.	$4x^2 - 3x - 7$	12.	$9a^2 - 18a + 8$
13.	$10x^2 - 23x + 12$	14.	$6x^2 + x - 1$

15. $2x^2 + 7x + 6$

16. $12x^2 + 11x - 5$

17. $20x^2 + 11x - 3$

18. $15x^2 + x - 2$

19. $6x^2 - 17x + 12$

THAT'S ALL FOLKS!!

DAY 26

A TIMEOUT!

Somehow our road trip stalled. It wasn't for the lack of gas or even the numerous mathematical rules. You guessed it correctly! Trying to drive a car, writing topics that can make math instructors pull out their hair (if any is left), and hitting the salient math facts that will make YOU a better student, the journey became mundane.

It is very difficult to typify the frustrated math instructor, standing in front of students, explaining a new concept. As instructors we daily explain a new topic, knowing full well that what was covered yesterday, two days, or even last week, will be forgotten or never understood by many students.

Understanding mathematical concepts requires dedication by both student and instructor. From the student, he or she must be willing to take careful notes, reexamine that new lecture prior to beginning the homework.

I'm sorry about the homework. Just like an accomplished pianist, stone mason, or clerical occupation, PRACTICE makes perfection. Math instructors call practice, HOMEWORK. Enough said!

As a math instructor, one needs to explain the day's concept so that the majority of the class has a clear unambiguous understanding of that day's topic. If the instructor fails to deliver the message, or the student neglects the PRACTICE, then disharmony in communication becomes apparent and FRUSTRATIONS materialize.

I am hoping both instructor and student can reach a happy medium. When and if that occurs, both sides of the table will be happy and enthusiastic.

Enough of the pontificating, let us return to math frustrations.

DAY 27

MORE FACTIONAL EQUATIONS!

I TRIED TO DISGUISE THE WORD "FRACTION" in "FRACTIONAL"but to no avail. That dreaded concept comes up all too frequently. Needless to say, we need to better equip our repertoire with a more complete understanding of fractional equations.

Several days ago, I explained how to solve relatively simple fractional equations. Now we need to dig deeper into this topic. SMILE!

Example 1: Solve: $\frac{4x+1}{7} - \frac{2x-1}{6} = 1\frac{1}{2}$

Solution: $\boxed{1\frac{1}{2} = \frac{3}{2}}$

Multiply by sides of the equation by the LCD, which is 42.

$$\frac{\overset{6}{\cancel{42}}(4x+1)}{\underset{1}{\cancel{7}}} - \frac{\overset{7}{\cancel{42}}(2x-1)}{\underset{1}{\cancel{6}}} = \frac{\overset{21}{\cancel{42}}(3)}{\underset{1}{\cancel{2}}}$$

$6(4x + 1) - 7(2x - 1) = 21(3)$
$24x + 6 - 14x + 7 = 63$

Combine like terms: $10x + 13 = 63$

Subtract 13 from both sides: $-13 \quad -13$

Divide both sides by 10: $\frac{10x}{10} = \frac{50}{10}$

The solution is: $x = 5$

Check it! This is a must!

Example 2: Solve: $\frac{7c-4}{c^2-c} = \frac{5}{c-1}$

Solution:

Cross-multiply. Remember from Day 19: $\frac{a}{b} = \frac{c}{d}$ is the same as: $ad = bc$.

$$\frac{7c-4}{c^2-c} = \frac{5}{c-1}$$

$$(7c-4)(c-1) = 5(c^2-c)$$

FOIL this side use distributive property on this side

Add 5c and subtract $5c^2$ from both sides.

$$\begin{array}{r} 7c^2 - 11c + 4 = 5c^2 - 5c \\ -5c^2 + 5c -5c^2 + 5c \\ \hline 2c^2 - 6c + 4 = 0 \end{array}$$

Factor out a 2, then factor $c^2 - 3c + 2$.

$$2(c^2 - 3c + 2) = 0$$
$$2(c-2)(c-1) = 0$$

Set all three factors equal to 0.

$2 = 0$	$c - 2 = 0$	$c - 1 = 0$
no	$c = 2$	$c = 1$

Two answers – check them out.

$c = 2$

$$\frac{7(2)-4}{(2)^2-2} = \frac{5}{2-1}$$

$$\frac{14-4}{4-2} = \frac{5}{1}$$

$$\frac{10}{2} = 5$$

$5 = 5$ ✓
2 is a valid solution

$c = 1$

$$\frac{7(1)-4}{(1)^2-1} = \frac{5}{1-1}$$

$$\frac{7-4}{0} = \frac{5}{0}$$

$$\frac{3}{0} = \frac{5}{0}$$

$$\frac{3}{0} \neq \frac{5}{0}$$

OOPS! Does not check!

If you have had Algebra II, you could have solved the previous problem. However, if you've had only Algebra I, you need to look at it this way:

$$\frac{7c-4}{c^2-c} = \frac{5}{c-1}$$

Factor the denominator of the fraction on the left:

$$\frac{7c-4}{c(c-1)} = \frac{5}{c-1}$$

Find the LCD for $(c^2 - c)$ and $(c - 1)$, which is $c(c - 1)$. Multiply both sides by the LCD:

$$\frac{c(c-1)(7c-4)}{c(c-1)} = \frac{c(c-1)(5)}{c-1}$$

$$7c - 4 = 5c$$

Subtract 5c and add 4 to both sides: $\frac{2c}{2} = \frac{4}{2}$

Divide both sides by 2: $c = 2$

Is it easier this way?

Try these for some PRACTICE!

1. $\frac{x}{4} + \frac{x}{3} = \frac{7}{12}$

2. $\frac{x}{3} - \frac{x}{4} = \frac{2}{3}$

3. $\frac{18}{x} - 4 = 2$

4. $\frac{x}{2} + \frac{x}{3} + \frac{x}{4} = 26$

5. $y + 1 - \frac{3}{4}y = \frac{1}{5}y$

6. $\frac{4}{y} - \frac{1}{2} = \frac{5}{12} - \frac{3}{2y}$

7. $\frac{15}{4y} + 2\frac{5}{12} = 3 - \frac{5}{y}$

8. $\frac{5}{8}x - \frac{1}{3}x = \frac{5}{6}x - 13$

9. $\frac{x+2}{4} - \frac{x-3}{3} = \frac{1}{2}$

10. $\frac{4}{3} + \frac{x-3}{4} = \frac{3x-1}{6}$

11. $\frac{x+4}{4} - \frac{3x-9}{7} = \frac{1}{2}$

12. $\frac{2+x}{6x} = \frac{3}{5x} + \frac{1}{30}$

13. $\frac{9}{2b+1}=3$

14. $\frac{10}{1-2m}=2$

15. $\frac{3x-1}{4x}=\frac{2x+3}{3x}$

16. $\frac{3}{5-3t}=\frac{1}{2}$

DAY 28

REMOVING PARENTHESES FIRST IN ORDER OF OPERATIONS

OHAHOA was our caps for Day 10. It pertained to right triangle trigonometry. Today we look at this idiom: My Dear Aunt Sally (MDAS). Multiply/divide, then add/subtract. In that order, from left to right, always. How about parentheses? It's easy. Remove them! That's why we start with them first. How is that done? Let's see.

Example 1: Evaluate $60 - 7(5 + 6 \div 2) + 2^4$

Solution:

Working within the parentheses, divide first:
$60 - 7(5 + 3) + 2^4$

Still in the parentheses, add: $60 - 7(8) + 2^4$

Next, the exponent: $60 - 7(8) + 16$

Multiply before add/subtract: $60 - 56 + 16$

Since subtraction comes first, do that: $4 + 16$

Then add: 20 Final answer.

Try a couple before we go further.

1. $33 - 3[4 \cdot (7 - 5)] + 3^2$
2. $74 - 4[3 \cdot (9 - 4)] + 5^2$

Once the parentheses are removed, operations are performed in this order:

(i) evaluate any powers first
(ii) next, multiply and divide, **IN ORDER,** from left to right
(iii) last, add and subtract, **IN ORDER,** from left to right

Let's try some more PRACTICE.

3. $16 + 4(7 - 4 + 1)$
4. $37 + 3[16 \div (2 \cdot 4)]$
5. $102 - 2(3^4 - 51)$
6. $17 + 3(8 - 5 + 1)$
7. $48 + 2[12 \div (2 \cdot 3)]$
8. $53 - 3(2^5 - 22)$

Now let us put some letters in our practice problems and see how it goes.

Example 2: $10d - (d - 2) =$

$10d - d + 2 =$
$9d + 2$

Example 3: $x^2 - (x - 3)(x - 2) =$

$x^2 - (x^2 - 5x + 6)$
distribute the negative sign
$x^2 - x^2 + 5x - 6 =$
$5x - 6$

Example 4: $5[20 + 3(n - 1)] =$

$5[20 + 3n - 3] =$
$5[20 - 3 + 3n] =$
$5[17 + 3n] =$
$85 + 15n$

Example 5: $5(r + 2) - 3(6 - r) =$

$5r + 10 - 18 + 3r =$
$5r + 3r + 10 - 18 =$
$8r - 8$

Try these for fun!

9. $-5x + (6 - 9x)$
10. $6(1 - 2x) + 5x$
11. $6x - (2x + 5)$
12. $3(x + 9) + 2(x - 6)$
13. $y^2 + (y + 3)(y - 5)$
14. $\frac{n}{2}[n^8 + 4(n - 2)]$
15. $4x^2 - 2(x + 1)(x + 4)$
16. $9m - 3(4 - m)$
17. $3y + 4(-8y + 7)$
18. $m^2 - (m - 6)(m - 2)$

That's all for today.

DAY 29

TRANSLATING WORDS INTO MATH SYMBOLS

FOREIGN LANGUAGE? NO! YES, IT IS math with words.

Oh, how dreadful. Nothing can be more exasperating than having to translate words into mathematical terminology.

WORDS	MATH SYMBOLS
Subtract 6 from 21, then add 3.	$21 - 6 + 3$
Subtract 6 from 21, then divide by 3.	$(21 - 6) \div 3$ or $\frac{21-6}{3}$
Add 3 to x, then square the result.	$(x + 3)^2$
The sum of the squares of x and 3.	$x^2 + 3^2$ or $x^2 + 9$
The quantity r minus t, cubed.	$(r - t)^3$
Four times the sum of 10 and z.	$4(10 + z)$
10 divided by the quantity r plus s.	$\frac{10}{r+s}$
Subtract the product of 4 and x from 12.	$12 - 4(x)$
Add 6 and a, then cube the result.	$(a + 6)^3$
Divide x by 7, then add y.	$\frac{x}{7} + y$
Divide the sum of a and b by the difference c minus d.	$\frac{a+b}{c-d}$

As you can observe by the examples and their conversion to math symbols, use the basic fundamentals (add, subtract, multiply, and divide) to translate into math symbols and make sure your "order of operations" is clearly defined. Use parentheses or powers when necessary. It takes **"PRACTICE"** to understand word translations. Remember the adage: "Practice helps to make it perfect!"

Try these and then take a well-deserved day off.

1. Multiply 8 by x and add y to the result.
2. 2 more than the product of 3 and d.
3. Eleven times the quantity y minus 3.
4. 5 divided by the quantity x plus 4.
5. Add the cubes of a and 2.
6. Subtract the square of n from 12.
7. The quantity 3x + 2 divided by the quantity 2 – 3x.
8. Add 5 and 4a, then square the result.
9. Add 6 to x which now equals 7.
10. Our solution 1 equals 3 + s.
11. Negative five plus the absolute value of p equals zero.
12. Negative four is greater than negative three.
13. Two squared plus 10 is less than four cubed.
14. Twenty-five divided by 5 added to 4 squared.
15. Forty-two equals six times x.
16. Six cubed is less than three hundred.

DAY 30

LCM VERSUS GCF

RAIN, AND I MEAN, HEAVY RAIN, dampened our quest to play another round of golf today. Waders and rowboats aren't too good for a golf course.

So we select another math concept that so many students misunderstand. That being **least common multiple** (LCM) which is often confused with **greatest common factor** (GCF) or greatest common divisor. What is confusing is the words least versus greatest.

LCM is looking for the smallest number divisible by all the numbers in question, whereas, the GCF is finding the largest number that is a factor (divisor) of all numbers in question. The LCM is always larger than the GCF of numbers in question. The LCM is always larger than the GCF of identical numbers.

Let's see how this can occur. Use the numbers 12 and 16.

$$12 \text{ factors into } 2^2 \cdot 3.$$
$$16 \text{ factors into } 2^4.$$

To find the LCM, after factoring the numbers into primes (check DAY 8 for a primes review), select every prime, raised to the largest power and then find their product. In 12 and 16, 2^4 and 3 are the primes raised to the largest power and that product is 48.

Using the same prime factors, now select any prime factor ***common*** to each factorization with the smallest exponent. Common to both 12 and 16 is 2^2 or 4, this is the GCF. Truly a small number compared to 48.

Let us practice by finding both the LCM and GCF of the following numbers.

1. 6 and 8
2. 5 and 10
3. 6 and 9
4. 2, 6, and 12
5. 10 and 8
6. 8 and 12
7. 4, 6, and 10
8. 2, 3, and 10
9. 2, 4, and 3
10. 10 and 4

DAY 31

LCM ONLY

WITHOUT ANY DELAY, WE NEED TO reinforce how to find the LCM with some hardy numbers. So briefly review Day 30 and its off to practice we shall go, we shall go, ...

Find the Least Common Multiple of the following numbers.

1. 42 and 56
2. 25 and 70
3. 16 and 42
4. 66 and 90
5. 60 and 108
6. 96 and 212
7. 72 and 120
8. 220 and 400
9. 42 and 24
10. 36 and 56
11. 8, 12, and 18
12. 36 and 144

DAY 32

GCF ONLY

AGAIN, WE NEED TO PRACTICE. THIS time, find the greatest common factor (GCF) in these numbers.

1. 42 and 56
2. 25 and 70
3. 16 and 42
4. 66 and 90
5. 60 and 108
6. 96 and 212
7. 72 and 120
8. 12, 60, and 90
9. 36 and 42
10. 24 and 66
11. 8, 36, and 54
12. 138 and 102
13. 24 and 42
14. 24 and 36

DAY 33

CHEATING TO FIND THE GCF

FRUSTRATIONS IN TEACHING MATHEMATICS ALWAYS SOLVES in unfriendly math problems. Sometimes we need to do it as instructed by your educator, and then learn how to cheat. So here it goes for GCF. First and foremost – get out the calculator.

Try this one: find the GCF of 840 and 3432.

Step 1: Divide 3432 by 840.

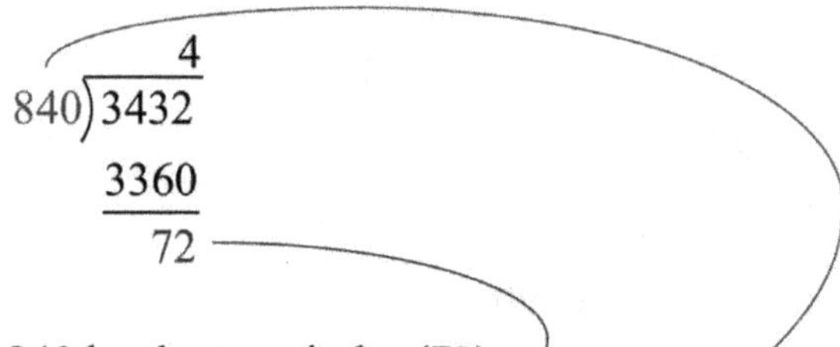

Step 2: Divide 840 by the remainder (72).

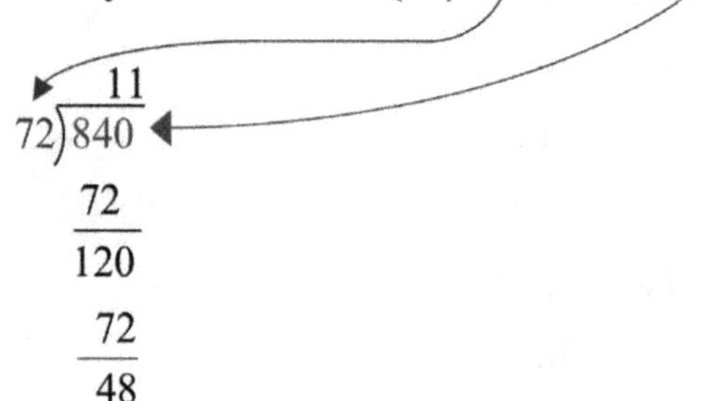

Step 3: Divide 72 by 48.

$$\begin{array}{r} 1 \\ 48\overline{)72} \\ \underline{48} \\ 24 \end{array}$$

Step 4: Divide 48 by 24.

$$\begin{array}{r} 2 \\ 24\overline{)48} \\ \underline{48} \\ 0 \end{array}$$

When we arrive at 0 for the remainder, the **last divisor** is the GCF! Further information about this can be obtained by referring to the topic "Euclidean Algorithm".

Now try this so you can cheat!

1. 24 and 54
2. 39 and 91
3. 72 and 160
4. 5291 and 11,951
5. 150 and 480
6. 72 and 120
7. 338 and 507
8. 1105 and 3289
9. 1421 and 2523
10. 484 and 363
11. 120 and 84
12. 546 and 390

DAY 34

WHY NOT CHEAT FOR THE LCM?

IF THERE IS A METHOD TO our madness, then there has to be a cheat for LCM.

Don't Worry!

Be Happy!

THERE IS ONE!

How does it work? Let's see. Find the LCM of 36 and 56.

Step 1: Identify the GCF of 36 and 56.

$$\begin{array}{r} 1 \\ 36\overline{)56} \\ \underline{36} \\ 20 \end{array} \qquad \begin{array}{r} 1 \\ 20\overline{)36} \\ \underline{20} \\ 16 \end{array} \qquad \begin{array}{r} 1 \\ 16\overline{)20} \\ \underline{16} \\ 4 \end{array} \qquad \begin{array}{r} 4 \\ 4\overline{)16} \\ \underline{16} \\ 0 \end{array}$$

So the GCF is 4.

Step 2: Multiply the two numbers and divide by the GCF.

$$\frac{36 \times 56}{4} = 504$$

The LCM is 504.

To summarize, LCM (36, 56) = $\frac{36 \times 56}{\text{GCF}}$. In general terms, LCM (a, b) = $\frac{a \bullet b}{\text{GCF}(a, b)}$.

Try these for some practice.

1. 42 and 56
2. 66 and 90
3. 25 and 70
4. 60 and 108
5. 16 and 42
6. 96 and 212
7. 72 and 120
8. 220 and 400
9. 36 and 48
10. 27 and 54
11. 22 and 35
12. 17 and 39
13. 24 and 36
14. 40 and 64
15. $6x^3y$ and $12x^2y^2$
16. 50 and 275

DAY 35

TRICKS FOR FINDING THE GCF AND LCM OF SEVERAL NUMBERS

LET US TRY ANOTHER METHOD IN obtaining the GCF and LCM for two numbers. They are 42 and 56.

Step 1: Find the prime(s) that divide into both numbers simultaneously. Perform the division. The product of the primes is the GCF.

Step 2: Multiply the GCF by the quotient (the bottom row) in the final division. There must NOT be any common factors in the bottom row.

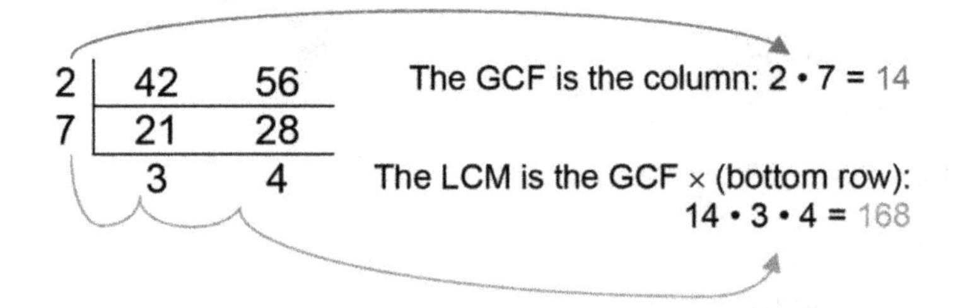

2	42	56
7	21	28
	3	4

The GCF is the column: 2 • 7 = 14

The LCM is the GCF × (bottom row):
14 • 3 • 4 = 168

Now try this one: 60 and 80

5	60	80
2	12	16
2	6	8
	3	4

The GCF is the column: 5 • 2 • 2 = 20

The LCM is the GCF × (bottom row):
20 • 3 • 4 = 240

And another: 66 and 90

2	66	90
3	33	30
	11	10

The GCF is the column: 2 • 3 = 6

The LCM is the GCF × (the bottom row):
6 • 11 • 10 = 660

Now try three: 24, 68, and 56

2	24	68	56
2	12	34	28
	6	17	14

The GCF is the column: 2 • 2 = 4

The LCM is the GCF × (bottom row):
4 • 6 • 17 • 14 = 5712

Practice. Find the GCF and LCM using this method.

1. 144 and 120
2. 147 and 70
3. 66 and 88
4. 12 and 25
5. 72 and 84
6. 45 and 80
7. 110 and 240
8. 235 and 180
9. 6, 9, and 12
10. 10, 20, and 40
11. 24, 30, and 48
12. 12, 24, and 60
13. 14, 35, and 70
14. 6, 18, and 24
15. 8, 20, and 36
16. 15, 60, and 75

DAY 36

COUNTING FACTORS

A PAUSE FROM LCM AND GCF is in order. We have been concentrating on these two concepts for several days and it is time to take a break.

Have you ever tried to figure out all the factors of a number? Most everyone says no. I just figured the factors of 12 are 1, 2, 3, 4, 6, and 12. It's real easy: (1)(12), (2)(6), and (3)(4). That's not hard. I agree.

However, let us try 48. Well, there is more. How many?? Don't bother me with details. I do know that 48 has lots. Let's figure out ***exactly*** how many factors 48 has.

48's prime factorization is $2^4 \times 3$. Examine the exponents of both 2 and 3. The exponent of 2 is 4 and the exponent of 3 is 1. Increase ***both*** exponents by 1 and then multiply. The 4 becomes 5 and the 1 becomes 2: $5 \times 2 = 10$. That's the number of factors for 48: 1, 2, 3, 4, 6, 8, 12, 16, 24, and 48.

Increase the exponents of the prime factors by 1 and then multiply the new numbers. PRESTO! That's the number of factor of the numeral.

Try these problems before you forget this new rule.

1. 144
2. 36
3. 25
4. 675
5. $1,234,567 = 5^2 \times 7^3 \times 11^4$
6. $456,533 = 7^3 \times 11^3$
7. 72
8. 100
9. 60
10. 250
11. 18
12. 24
13. 64

DAY 37

SO YOU THOUGHT THAT FACTORING WAS A BREEZE!

1. Factor completely.

   a. $6x^2 - 6x - 36$ b. $x^5 + 2x^4 - 24x^3$

2. Find the pair of numbers whose product is 11 and whose sum is 12.

   Use the FOIL method to show that $(9x + 18)(x - 10)$ is $9x^2 - 72x - 180$. If you where asked to completely factor $9x^2 - 72x - 180$, why would it be incorrect to give $(9x + 18)(x - 10)$ as your answer?

3. What steps would you take to factor $x^2 - 9x + 20$?
4. Factor completely: $x^2 - x - 56$
5. Find the pair of numbers whose product is 24 and whose sum is 11.
6. Factor completely.

   a. $x^2 + 2x - 120$

   b. $3x^3 + 9x^2y - 30xy^2$

c. $2x^2 - 10x + 12$

7. In factoring a trinomial in "z" as $(z + a)(z + b)$, what must be true of "a" and "b", if the coefficient of the last term of the trinomial is positive?

8. Factor completely.

   a. $x^2 + 2xy - 99y^2$

   b. $5x^3 + 10x^2 - 75x$

   c. $x^2 - x - 45$

9. Explain the error in the following:

   $x^2 + 2x - 15 = (x - 5)(x + 3)$

10. Factor completely.

    a. $x^2 - 8x - 48$ b. $x^2 + 14x - 15$

11. Complete the factoring.

    a. $x^2 - 2x - 8 = (x + 2)(\quad)$

    b. $x^2 + 12x + 35 = (x + 7)(\quad)$

12. Factor completely: $x^2 + 3xy - 10y^2$

    Factor as completely as possible. If unfactorable, indicate that the polynomial is prime.

13. $12x^2 + 17x + 6$

    a. $(12x + 2)(x + 3)$b. $(3x - 2)(4x - 3)$

    c. $(3x + 2)(4x + 3)$d. prime

14. $6x^2 - 5xt - 6t^2$

    a. prime b. $(3x + 2t)(2x - 3t)$

    c. $(6x + 2t)(x - 3t)$ d. $(3x - 2t)(2x + 3t)$

15. $56 - 15x + x^2$

    a. $(x - 7)(x - 8)$ b. $(x + 7)(x + 8)$

    c. $(x + 7)(x - 8)$ d. $(x - 7)(x + 8)$

16. $15z^2 - 2z - 8$

    a. $(15z + 2)(z - 4)$b. $(3z - 2)(5z + 4)$

    c. prime d. $(3z + 2)(5z - 4)$

17. $12x^2 - 5xt - 3t^2$

    a. $(3x - t)(4x + 3t)$ b. $(3x + t)(4x - 3t)$

    c. $(12x + t)(x - 3t)$ d. prime

DAY 38

GAME DAY

WE NEED TO TRY A FEW math games. I'll provide an example with several problems to try. Good luck and put your game-day smile on so you can beat the odds.

Example 1: Division Problem with 7 given numbers. Find the missing 17 numbers.

			1	□	□
□	□	5	□	□	□
		3	6		
		□	7	2	
		□	□	□	
			□	□	□
			□	□	□
					0

Solution:

			1	4	8
3	6	5	3	2	8
		3	6		
		1	7	2	
		1	4	4	
			2	8	8
			2	8	8
					0

Practice 1a: Division Problem with 7 given numbers. Find the missing 17 numbers.

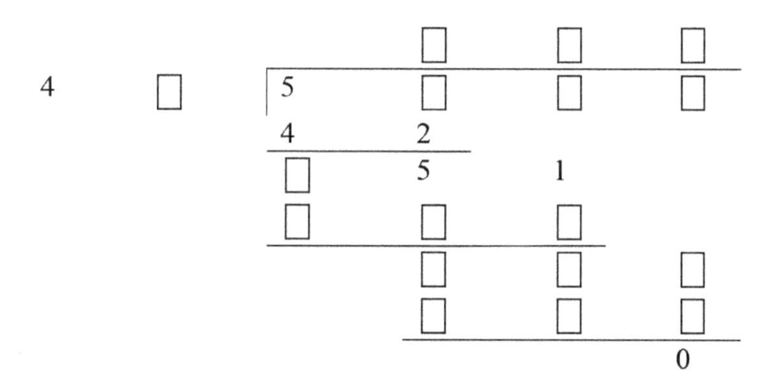

Solution:

			1	3	6
4	2	5	7	1	2
		4	2		
		1	5	1	
		1	2	6	
			2	5	2
			2	5	2
					0

Practice 1b: Seven's and one's. Find the missing 14 numbers.

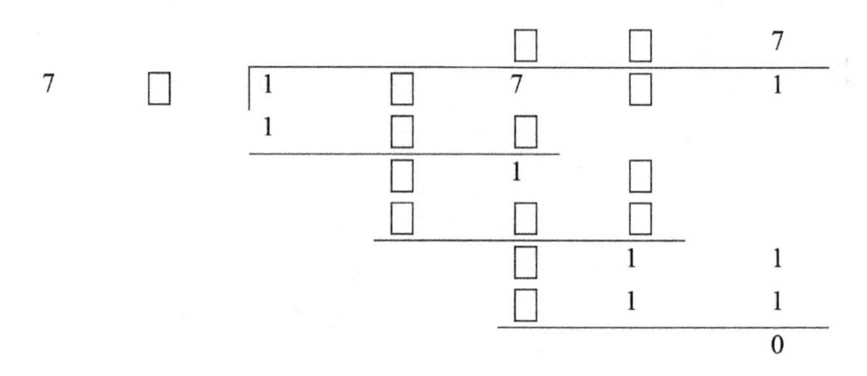

Solution:

				2	5	7
7	3	1	8	7	6	1
		1	4	6		
			4	1	6	
			3	6	5	
				5	1	1
				5	1	1
						0

Practice 1c: Division Problem with 12 given numbers. Find the missing 20 numbers.

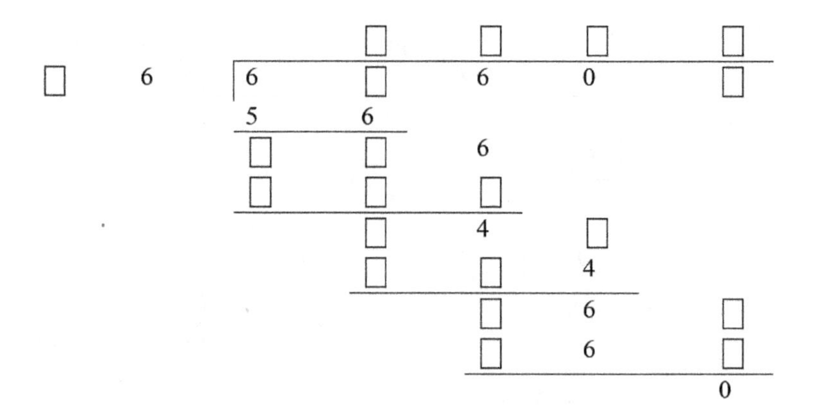

Solution:

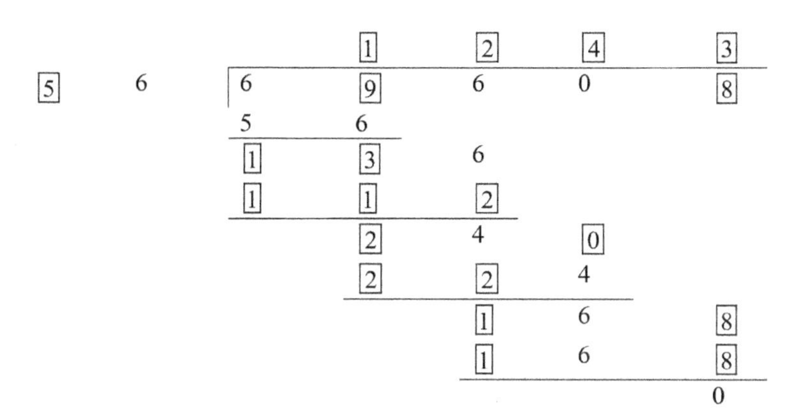

Practice 1d: Fill in the boxes.

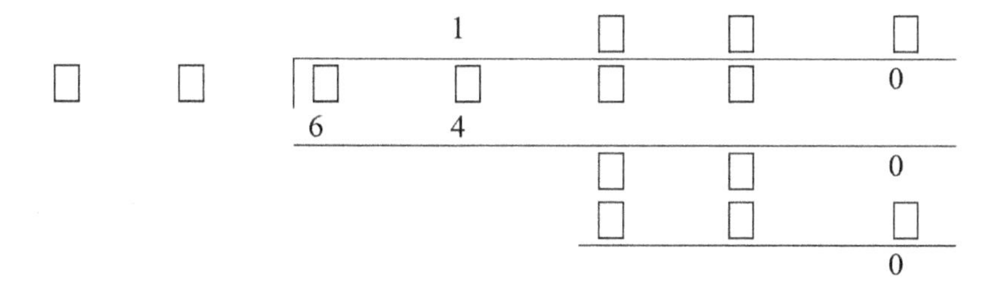

Solution:

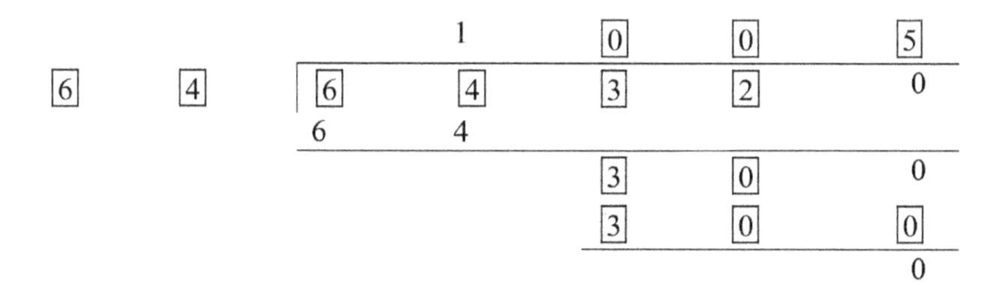

Example 2: Magic Square. Place the digits 1 through 9 into a 3 × 3 square so that each row, column, and diagonal add up to the same total. Use the digits 1 through 9 only once.

Solution:

	3	
9		
2		6

4	3	**8**
9	**5**	**1**
2	**7**	6

Practice 2a:

		4
7		
	1	8

Solution:

2	**9**	4
7	**5**	**3**
6	1	8

Practice 2b:

	1	
3		
4	9	

Solution:

8	1	**6**
3	**5**	**7**
4	9	**2**

Practice 2c:

6		2
	5	
	3	

Solution:

6	**7**	2
1	5	**9**
8	3	**4**

Example 3: Counting Squares. How many squares in this figure?

a	b
c	d

5 squares
a, b, c, d, and (abcd)

Practice 3a:

a	b	c
d	e	f
g	h	i

14 squares

a, b, c, d, e, f, g, h, i, abde, bcef, degh, efhi, abcdefghi

Practice 3b:

a	b	c	d	e
f	g	h	i	j
k	l	m	n	o
p	q	r	s	t
u	v	w	x	y

Practice 3c:

a	b	c	d
e	f	g	h
i	j	k	l
m	n	o	p

DAY 39

INEQUALITIES

YOU'VE HEARD THE EXPRESSIONS "AT MOST" or "at least" or "no more than" or "is no less than". What do they **really** mean in mathematical terminology.

You guessed it. The thunder-boomers with lightning raced across the sky and left me contemplating what is today's topic. The course was closed and surely wouldn't open today with this torrential downpour. The sand traps would look like a pond in the middle of a pasture of manicured grassland.

How did "inequalities" come into play. I guess it came from tee to green at the golf course. The long hitters are set further away from the green, while short hitters get to hit their golf shot closer to the pin. That's an inequality.

Preferential treatment, you say. Not really. Just one of the many handicaps duffers face when trying to explain their golf stories.

Enough about golf. What do the opening expressions truly represent?

a. "at least" means no less than
   → uses the symbol: $\geq$

b. "at most" means no more than
   → uses the symbol: $\leq$

c. "no more than" means at most
→ uses the symbol: $\leq$

d. "no less than" means at least
→ uses the symbol: $\geq$

e. "x" is between 4 and 8
→ $4 < x < 8$

Examples:

y is at least 7 → $y \geq 7$
x is at most 4 → $x \leq 4$
7 is no more than 10 → $7 \leq 10$
a is no less than 15 → $a \geq 16$

Practice:

1. "a" is less than 6
2. "b" is greater than 12
3. "c" is no more than 20
4. "d" is between 3 and 5
5. "e" is no more than 13
6. "f" is between 5 and 8
7. "g" is at least 9
8. "h" is between 6 and 12
9. "j" is no less than 3
10. "k" is at most 19
11. "l" is at least 15
12. "m" is no less than 40
13. "n" is between 2 and 7
14. "p" is at least 23

15. "q" is between 10 and 15
16. "r" is at most 52
17. "s" is between 7 and 15
18. "t" is no less than 33
19. "u" is at least 17
20. "v" is at most 29
21. "w" is no more than 18
22. "x" is between 5 and 9
23. "y" is no less than 47
24. "z" is no more than 27

DAY 40

LAST TRY → FACTORING

Factor the following trinomials. There are NO primes.

1. $u^2 - 3uv - 54v^2$
2. $u^2 - 5uv - 14v^2$
3. $s^2t^2 - 15st + 26$
4. $x^2y^2 + 7xy + 12$
5. $-m^2 + 6m + 40$
6. $x^2 + 3xy - 28y^2$
7. $x^2 - x - 42$
8. $x^2 + 4xy - 192y^2$
9. $u^2 - 4uv - 45v^2$
10. $u^2 - 2uv - 15v^2$
11. $12x^2 + 17x + 6$
12. $15y^2 + 26y + 8$
13. $12z^2 - 7z - 12$
14. $8z^2 + 6z - 9$
15. $24m^2n^2 - 79mn + 40$

16. $28m^2 + 81mn + 56n^2$
17. $15x^2 + 26x + 8$
18. $6y^2 + 17y + 12$
19. $12z^2 + 7z - 12$
20. $20z^2 + 7z - 6$

DAY 41

MATHEMATICAL SKETCHING

We are changing our direction 180° today and try our skill at mathematical sketching. Learning how to draw a line or plot points on a grid will enable you to move quickly through coordinate geometry in your algebra study.

A quick review of the basics follows:

1. The general equation of a line is **Ax + By = C**. The slope-intercept form of a line is **y = mx + b,** where "m" is the slope and "b" is the y-intercept.
2. Positive and negative values of the x and y coordinates in a Cartesian Coordinate System.

   x is positive in the 1st and 4th quadrants.

   y is positive in the 1st and 2nd quadrants.

Cartesian Coordinate System

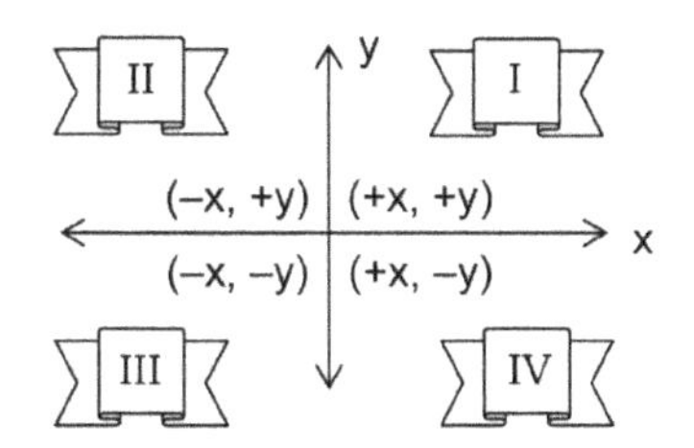

3. Graphing a linear equation. Convert the equation to slope-intercept form:

$$y = mx + b.$$

Example 1: $y = 4x - 3$

Solution: Using $m = 4$ and $b = -3$, graph the equation.

Step 1: Draw a grid.

Step 2: Label the x- and y-intercepts.

Step 3: Locate the y-intercept, –3 ("b") and label it. This is where the line crosses the y-axis.

Step 4: The slope, "m", is 4. "m" means $\frac{\Delta y}{\Delta x}$ or $\frac{\text{change in y}}{\text{change in x}}$ or $\frac{4}{1}$ where Δ is "delta".

For every +4 units in y there is a +1 unit change in x. Starting from the y-intercept, –3, move up +4 units, landing at +1. Now move +1 on the x-axis. This is where there is another point on the line.

Step 5: Connect the two points (–3, 0) and (1, 1) and you have created the line that satisfies the equation $y = 4x - 3$.

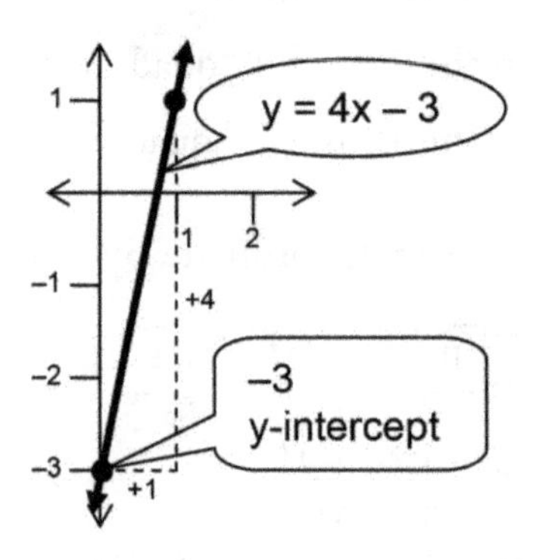

Example 2: $2x + 5y = 16$

Solution:

Step 1: Solve for y: $5y = -2x + 16$

$$y = -\frac{2}{5}x + \frac{16}{5}$$

Using $m = -\frac{2}{5}$ and $b = \frac{16}{5}$, graph the original equation.

Step 2: Draw a grid.

Step 3: Locate the y-intercept, "b". This is where the line crosses the y-axis.

Step 4: Play with the slope, "m".

$$-\frac{2}{5} \text{ means } \frac{\Delta y}{\Delta x} \text{ or } \frac{\text{change in y}}{\text{change in x}}$$

For every –2 units on the y-axis there is a +f units movement on the x-axis. Now you have located another point on the grid! Two points determine a line. Connect the two points!

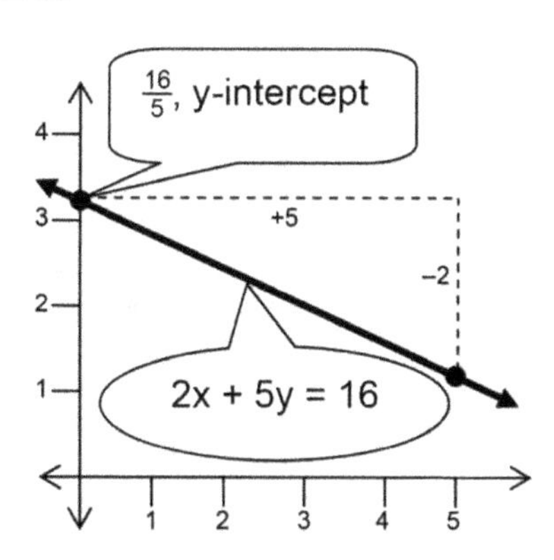

4. Graphing an ordered pair.

Example 3: Graph the ordered pair: (3, 7)

Example 4: Graph the ordered pair: (–3, 2)

Example 5: Graph the ordered pair: (4, –1)

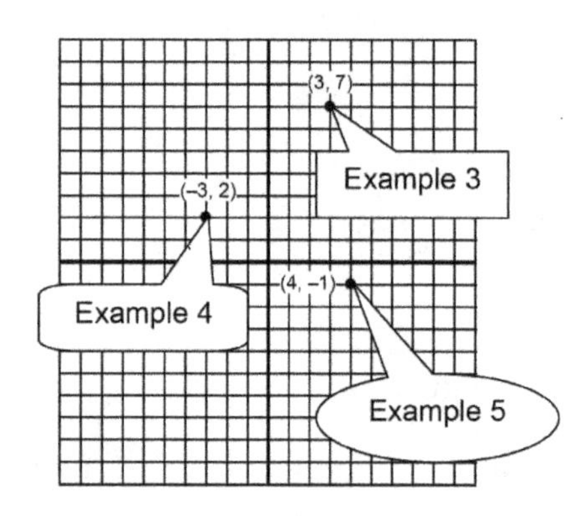

5. Given a point, is it a solution to the given equation?

Example 6: Is (2, 3) a solution of the equation 3x – 2y = 7?

Solution: Substitute the point into the equation.

$$3(2)-2(3)\overset{?}{=}7$$

$$6-6\overset{?}{=}7$$

$$0\neq 7$$

∴ the point (2, 3) is ***NOT*** *a* solution for 3x – 2y = 7

Graph these equations to see if you have mastered today's lesson. What is the slope and y-intercept of each equation?

1. y = 2x + 3   2. y = –3x + 1

Try these ordered pairs to see if it is a solution for the equation 4x + 2y = 18.

3. (2, 5)
4. (1, 8)
5. $(\frac{1}{2}, 8)$
6. (–5, 18)

DAY 42

GRAPHING A LINEAR EQUATION

THERE ARE THREE DISTINCT WAYS TO graph a line. Let us explore each one.

A. Slope and y-intercept.

Let us start with slope and y-intercept. At the conclusion of Day 41, we tried graphing several equations to get the hang of it. Now let us try it again.

Example 1: Slope of −2 and y-intercept of +2.

Solution:

Step 1: Draw a grid.

Step 2: Locate the y-intercept, +2.

Step 3: Starting from the y-intercept location, work the slope into the problem:

$$\frac{\Delta y}{\Delta x} = \frac{-2}{+1} \text{ down 2, right 1}$$

Step 4: Connect the two points and our line is created!

By the way, the equation of this line is $y = -2x + 2$.

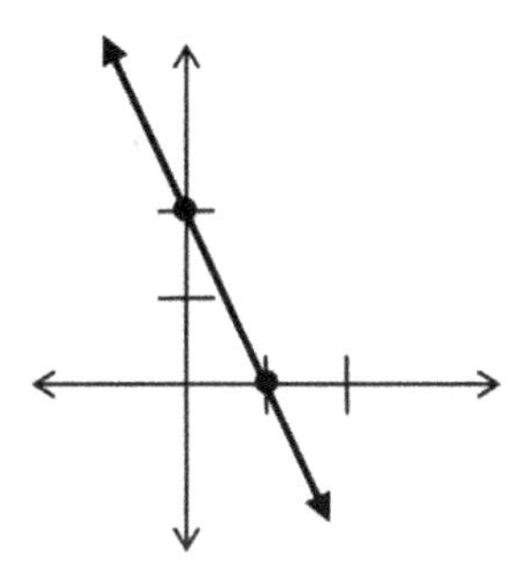

Example 2: If the slope, "m", of an equation is 2 and the y-intercept if 4, what is the equation of the line?

Solution:

Step 1: Substitute "m" and "b" into the equation $y = mx + b$.

$m = 2$ $\qquad$ $b = 4$

$$y = mx + b$$

$$y = 2x + 4$$

Step 2: Put it in standard form, $Ax + By = C$.

$$Ax + By = C$$

$$\left.\begin{array}{c} 2x - y = -4 \\ \text{or} \\ -2x + y = 4 \end{array}\right\} \text{either is okay}$$

B. One point and slope.

Example 3: If the slope ("m") is $\frac{3}{4}$ and the point is $(-4, -1)$, find the equation that satisfy these conditions.

$y - y_1 = m(x - x_1)$	where m = slope and (x_1, y_1) is a point
$y - (-1) = \frac{3}{4}(x - (-4))$	first rid the negative out of the equation
$y + 1 = \frac{3}{4}(x + 4)$	multiply everything by 4
$4(y + 1) = 4[\frac{3}{4}(x + 4)]$	remember on the right side that the 4's cancel
$4y + 4 = 3(x + 4)$	expand the right side
$4y + 4 = 3x + 12$	rewrite in standard form
$4y - 3x = 8$ or $3x - 4y = -8$	either equation is okay!

Find the equation of the following. Answers should be in standard form and no fractions in the answer.

1. (2, 7), m = 3 2. (–4, 1), m =

C. Two Points.

$$\text{Remember: } m = \frac{\text{change in y}}{\text{change in x}} = \frac{\Delta y}{\Delta x}$$

Example 4: Given two points (2, 6) and (7, 15), find the equation of the line.

Solution:

Step 1: Find the slope.

$$(x_1, y_1) = (2, 6)$$
$$(x_2, y_2) = (7, 15)$$

$$m = \frac{\Delta y}{\Delta x} = \frac{y_2 - y_1}{x_2 - x_1} = \frac{15 - 6}{7 - 2} = \frac{9}{5}$$

Step 2: Substitute the point (2, 6) and the slope $m = \frac{9}{5}$ into the equation

$$y - y_1 = m(x - x_1).$$

$$y - 6 = \frac{9}{5}(x - 2)$$

$$y - 6 = \frac{9}{5}x - \frac{18}{5}$$

$$5(y) - 5(6) = 5\left(\frac{9}{5}x\right) - 5\left(\frac{18}{5}\right)$$

Step 3: Expand the right side.

$$5y - 30 = 9x - 18$$

Step 4: Get a common denominator, which is 5. Multiply it times ***all*** four terms of the equation.

Step 5: Collect like terms to arrive at the equation of the line in standard form.

5y	−	30	=	9x	−	18
	+	30			+	30
		5y	=	9x	+	12
−9x				−9x		
−9x	+	5y	=	12		

The equation of the line in standard form is

−9x + 5y = 12

Let's try another one.

Example 5: Given the two points (4, 10) and (6, 12), find the equation of the line.

Solution:

Step 1: Find the slope.

$$m = \frac{\Delta y}{\Delta x} = \frac{y_2 - y_1}{x_2 - x_1} =$$

$$\frac{12-10}{6-4} = \frac{2}{2} = 1$$

Step 2: Using the point (4, 10) and m = 1

or

using the point (6, 12) and m = 1

(4, 10) and m = 1	(6, 12) and m = 1
$y - 10 = 1(x - 4)$ $y - 10 = x - 4$	$y - 12 = 1(x - 6)$ $y - 12 = x - 6$
$-6 = x - y$ **or** **$-x + y = 6$**	**$-6 = x - y$** **or** **$-x + y = 6$**

As you can see, it doesn't matter which point you use, you will get the same equation!

Try these problems. Remember, answer in standard form and no fractions in the answer.

3. (–4, 0) and (0, 2)

4. $(\frac{1}{2}, \frac{1}{3})$ and $(-\frac{1}{4}, \frac{5}{4})$

5. $(-\frac{2}{3}, \frac{8}{3})$ and $(\frac{1}{3}, \frac{7}{3})$

Given this information, find the equation of a line, in standard form, with no fractions in the answer.

6. (–1, 7), (–2, 11)
7. (4, –3), (–5, 3)
8. slope is –1; point (–5, 6)
9. (–9, 5), (11, 5)
10. slope is 4; point (–2, –3)
11. slope is 1; y intercept is 18
12. slope is 0; point (3, 7)

DAY 43

MORE IDEAS ABOUT LINEAR EQUATIONS

FIND THE X- AND Y-INTERCEPTS OF an equation. They are sometimes referred to as the ***zeros*** of the equation.

Example 1: In the equation, $2x + 3y = 12$, find the values of x and y respectively if one of the ordered pair is zero.

Solution:

When $x = 0$, then $3y = 12$ or $y = 4$.
Ordered pair: $(0, 4) \leftarrow$ y-intercept

When $y = 0$, then $2x = 12$ or $x = 6$
Ordered pair: $(6, 0) \leftarrow$ x-intercept

Example 2: Find the x- and y-intercepts of $2x - 3y = 18$.

Solution:

When $x = 0, y = -6$. When $y = 0, x = 9$.

$$\left.\begin{matrix}(0, -6)\\(9, 0)\end{matrix}\right\}$$ Two points. Now you can graph it!

Vertical and Horizontal Lines.

Vertical Lines $x = k \rightarrow$ parallel to the ***y-axis***.
Horizontal Lines $y = k \rightarrow$ parallel to the ***x-axis***.

Remember it!!
Vertical (V) or Horizontal (H) are ***always*** parallel to the opposite variable.

Example 3: Draw y = 2 and x = 5 on the same grid.

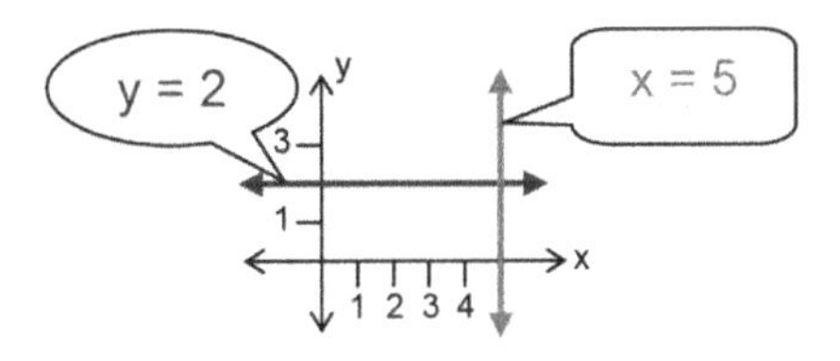

Slope of a Line Through Two Points.

Example 4: What is the slope of the line through the points (–3, 5) and (–4, –7)?

Solution:

$$m = \frac{\Delta y}{\Delta x} = \frac{y_2 - y_1}{x_2 - x_1} = \frac{-7-5}{-4-(-3)} = \frac{-12}{-1} = \frac{12}{1} \text{ or } 12$$

Remember to watch out for vertical and horizontal lines since the slopes are undefined and zero (0) respectively.

Now try these. Find the slope.

1. $(-4, -5)$ and $(-5, -8)$
2. $(6, -5)$ and $(-12, -5)$
3. $(-8, 6)$ and $(-8, -1)$
4. What would be the equation of problem 2?
5. What would be the equation of problem 3?

DAY 44

PARALLEL, PERPENDICULAR, OR NEITHER LINES

By definition, parallel lines have the ***same slope***.

Example 1: $\left.\begin{array}{l}\text{slope of line 1 is 2}\\ \text{slope of line 2 is 2}\end{array}\right\}$

lines are parallel

Perpendicular lines ***form right angles***. How do you find perpendicular lines? The product of the slopes of BOTH lines equals −1.

$$m_1 \cdot m_2 = -1$$

Example 2: $\left.\begin{array}{l}\text{slope of line 1 is } \dfrac{3}{5}\\ \text{slope of line 2 is } -\dfrac{5}{3}\end{array}\right\}$

lines are perpendicular

$$\frac{3}{5} \bullet -\frac{5}{3} = -1$$

Example 3: slope of line 1 is $\frac{2}{3}$, slope of line 2 is $-\frac{3}{2}$
lines are perpendicular

Lines, in the same plane, that are not parallel nor perpendicular are ***intersecting*** (neither parallel or perpendicular).

Example 4: slope of line 1 is $-\frac{1}{2}$, slope of line 2 is 1
lines are neither parallel or perpendicular

Now let's get ready to rumble and try real equations. Are these pairs of lines parallel, perpendicular, or neither?

1. $2x + 5y = 4$
   $4x + 10y = 1$
2. $-4x + 3y = 4$
   $-8x + 6y = 0$
3. $8x - 9y = 6$
   $8x + 6y = -5$
4. $3x - 2y = 6$
   $2x + 3y = 3$
5. $y = 2x - 3$
   $2y = -2x + 4$
6. $5x - 3y = -2$
   $3x - 5y = -8$
7. $5x + 3y = 2$
   $3x - 5y = -1$

Taking our understanding of linear equations one step further, we will try some harder problems.

Example 4: Write an equation, in standard form, passing through the point $(-2, 3)$ and parallel to $2x - 4y = 5$.

Solution:

Step 1: Find the slope. Remembering that $y = mx + b$, where "m" is the slope, solve the equation $2x - 4y = 5$ for y.

$$2x - 4y = 5$$
$$-4y = 5 - 2x$$
$$4y = 2x - 5$$
$$y = \frac{1}{2}x - \frac{5}{4}$$

Step 2: Use the slope, $m = \frac{1}{2}$, and the point-slope formula, $y - y_1 = m(x - x_1)$, to obtain our desired linear equation.

$$y - 3 = \frac{1}{2}(x - (-2))$$
$$y - 3 = \frac{1}{2}(x + 2)$$
$$y - 3 = \frac{1}{2}x + 1$$

Step 3: Rewrite in standard form.

$$2(y-3)=2(\frac{1}{2}x+1)$$
$$2y-6=x+1$$
$$2y=x+7$$
$$\mathbf{-x+2y=7 \text{ or}}$$
$$\mathbf{x-2y=-7}$$

$\mathbf{x-2y=-7}$ is the equation of a line, in standard form, passing through the point $(-2, 3)$ and parallel to $2x - 4y = 5$

Example 5: Write and equation, in standard form, passing through the point $(2, 3)$ and perpendicular to $3x - 4y = 8$.

Solution:

Step 1: Find the slope of the line $3x - 4y = 8$.

$$3x-4y=8$$
$$-4y=-3x+8$$
$$4y=3x-8$$
$$y=\frac{3}{4}x-2$$

Step 2: Find the slope of a line perpendicular ($\perp$) to $3x - 4y = 8$.

$$m_1 \bullet m_2=-1$$
$$\frac{3}{4} \bullet m_2=-1$$
$$\frac{4}{3}\bullet\frac{3}{4}\bullet m_2=-1\bullet\frac{4}{3}$$
$$m_2=-\frac{4}{3}$$

Step 3: Use the slope, $m = -\frac{4}{3}$, and the point-slope formula, $y - y_1 = m(x - x_1)$, to obtain our desired linear equation.

$$y - 3 = -\frac{4}{3}(x - 2)$$
$$y - 3 = -\frac{4}{3}x + \frac{8}{3}$$

Step 4: Rewrite in standard form.

$$3(y - 3) = 3(-\frac{4}{3}x + \frac{8}{3})$$
$$3y - 9 = -4x + 8$$
$$\mathbf{4x + 3y = 17}$$

4x + 3y = 17 is the equation of a line, in standard form, passing through the point (2, 3) and perpendicular to $3x - 4y = 8$.

Now try these seven:

8. Write an equation in standard form passing through the point (2, –3) and parallel to $3x - 4y = 5$.
9. Write an equation in standard form passing through the point (–1, 4) and perpendicular to $2x + 3y = 8$.
10. Write an equation in standard form of a line through $(-\frac{1}{2}, -\frac{1}{2}$ and perpendicular to $x = -3$
11. Write an equation in standard form that passes through the point (–1, 3) and is parallel to the graph of $3x - y = 4$.
12. Write an equation in standard form that has the same slope as $x - y = -4$ and containing (–4, 0) point.

13. Write an equation in standard form of a vertical line containing (3, 2).
14. Write an equation in standard form parallel to $2x - 3y = 12$, with the same y intercept as $x + 2y = 18$.

DAY 45

CALCULATORS, PRACTICE, AND DON'T GIVE UP

As our road trip heads towards the finish line, we need to examine several topics that may help you become an excellent mathematics student.

CALCULATORS

A calculator is a necessity for any and all if you desire to excel in higher math courses. However, the basics need to be learned without the aide of hand-held devices, such as calculators (non-graphic as well as graphic), computers, or space-age electronics. I will reiterate that knowing the multiplication table, the 10×10 grid located on Day 7 is the requirement for success. The days of using your fingers are long gone. Once the multiplication table is mastered, an electronic gadget may be required to simplify calculations. Read the instructions manual, never throw the booklet away. And always have it handy when doing homework.

PRACTICE

Most athletic teams, musical productions, theatrical plays, concerts, and shows are successful because the participants hone their innate skills by practice, practice, practice! One does not make the previously-mentioned activity without many hours of practice. This practice prior to the tryouts and cast or team selections, requires a lot of dedication to succeed.

Whether it is in middle school math (working with fractions and decimals), junior high school math (learning to understand the math alphabet), algebra, trigonometry, calculus, or statistics, practice in math skills will make you, the learner, more qualified to be a top-notch student. Sometimes it takes more time and effort to complete the course.

Homework, which is practice, makes YOU a well-rounded student. Strive to obtain competency in any math course or, for the matter, in each and every subject.

PRACTICE HELPS TO MAKE PERFECT(ION)!! DON'T GIVE UP!!

Success comes at a price. Put all your effort into a course or subject and make certain you find the time. Remember, education is your primary goal while in school. Ask questions when the topic is unclear. RAISE YOUR HAND for clarification or help. You are the determining factor, prior to any exam, in knowing if you understand the concepts being taught. Don't leave any stone unturned in accomplishing your goal. Golfers, artists, and musicians, all practice many hours prior to their performance. That is what you are doing when homework calls. Give it your best shot and success will follow!

ANSWERS

DAY 6

1. 10,000 **2.** 100,000 **3.** 1,000,000 **4.** 10,000,000 **5.** 100,000,000 **6.** six thousand four hundred twenty-one **7.** three hundred seventy-two thousand, six hundred twelve **8.** four million, five hundred six thousand, seventy-five **9.** thirty-seven thousand, four hundred two **10.** six hundred seventy-two million, three hundred forty-five thousand, nine hundred eighty-one **11.** 63,212,557 **12.** 89,739 **13.** 921,673 **14.** 234,569,004 **15.** 4070

DAY 10

1. 46° 66' **2. a.** 53° 13' **b.** 36° 47'

DAY 15

1. $\frac{4}{5}+\frac{2}{3}=\frac{12}{15}+\frac{10}{15}=\frac{22}{15}=1\frac{7}{15}$

2. $3\frac{3}{4}+1\frac{7}{10}=3\frac{15}{20}+1\frac{14}{20}=4\frac{29}{20}=5\frac{9}{20}$

3. $5\frac{2}{7}-2\frac{9}{14}=5\frac{4}{14}-2\frac{9}{14}=4\frac{18}{14}-2\frac{9}{14}=2\frac{9}{14}$

4. $2\frac{5}{6}+1\frac{7}{10}+3\frac{2}{5}=2\frac{25}{30}+1\frac{21}{30}+3\frac{12}{30}=6\frac{58}{30}=7\frac{28}{30}=7\frac{14}{15}$

5. $4\frac{1}{2}-3\frac{2}{3}=4\frac{3}{6}-3\frac{4}{6}=3\frac{9}{6}-3\frac{4}{6}=\frac{5}{6}$

6. $7\frac{5}{12}-3\frac{2}{3}=7\frac{5}{12}-3\frac{8}{12}=6\frac{17}{12}-3\frac{8}{12}=3\frac{9}{12}=3\frac{3}{4}$

DAY 16

1. \$1.13 **2.** 495 **3.** 1.2% **4.** 250 **5.** 300 **6.** 25% **7.** 40 **8.** 2% **9.** 27 **10.** 12½% **11.** 30 **12.** 33⅓% **13.** 125% **14.** 5.6 **15.** 520

DAY 17

1. $(2 + c)(a + b + x)$ **2.** $(y + 2)(x^2 - 5)$ **3.** $(x - 2)(x - 6)$ **4.** $(p + 5)(q + 2)$ **5.** $(2x + 3)(y + 1)$ **6.** $(a - 2)(2a + 3b)$ **7.** $(x + 3)(x^2 - 5)$ **8.** $(a + 4)(6x + 1)$ **9.** $(x - 5)(2x + 3y)$ **10.** $(p + q)(2 + a)$ **11.** $(x^2 + 4)(2x + 1)$ **12.** $(a^2 - 5)(2a - 1)$ **13.** $(2x + 7z)(2x - y)$ **14.** $(n - 4)(7t + 1)$ 15. $(x + 4y)(x - 4y + 2)$

DAY 18

1. 3¾ hours **2.** 48 hours **3.** 13⅓ min **4.** 5⅓ hours

DAY 19

1. $\frac{5 \text{ hours}}{260 \text{ miles}} = \frac{x \text{ hours}}{468 \text{ miles}} \rightarrow (5)(468) = 260x \rightarrow 2340 = 260x \rightarrow x = 9$ hours.

2. $\frac{20-\text{ft tree}}{5-\text{ft shadow}} = \frac{x-\text{ft building}}{15-\text{ft shadow}} \rightarrow 5x = (20)(15) \rightarrow 5x = 300 \rightarrow x = 60$-ft tall building

3. $\frac{\frac{1}{8}}{4} = \frac{2}{x} \rightarrow \frac{x}{8} = 8 \rightarrow x = 64$ miles

4. $\frac{7}{5} = \frac{2134}{x} \rightarrow 7x = (2135)(5) \rightarrow 7x = 10{,}675 \rightarrow x = 1525$ people.

5. $\frac{4}{22} = \frac{x}{50} \rightarrow 22x = (4)(50) \rightarrow 22x = 200 \rightarrow x = 9\frac{2}{22} \rightarrow$

You can buy 9 CDs with \$50, since you can't but $\frac{2}{22}$ of a CD.

DAY 20

1. $3y(2y + 3)^2$ **2.** $(5a - 2b - 1)(5a + 2b + 1)$ **3.** $(e + 1)(e - 2)$ **4.** $(3z)(3z + 1)(3z + 1)$ **5.** $(2c - 3)(3c + 5)$

DAY 22

1. $(a + 3)(a - 5)$ **2.** $(x + 9)(x - 4)$ **3.** $(b - 3)(b + 7)$ **4.** $(x - 7)(x + 5)$ **5.** $(x - 2)(x + 13)$ **6.** $(c + 8)(c - 5)$ **7.** $(r + 12)(r - 8)$ **8.** $(m + 9)(m - 6)$ **9.** $(y - 4)(y + 7)$ **10.** $(e + 9)(e - 11)$ **11.** $(f + 16)(f - 5)$ **12.** $(a - 12)(a + 3)$ **13.** $(x - 8)(x + 6)$ **14.** $(x - 10)(x + 5)$ **15.** $(x - 20)(x + 3)$

DAY 24

1. $x = 21$ **2.** $y = -\frac{3}{2}$ **3.** $x = 10$ **4.** $x = -6$ **5.** $x = -10$ **6.** $x = -35$ **7.** $x = 2\frac{7}{12}$ **8.** $x = 7\frac{1}{6}$ **9.** $x = 6$ **10.** $x = -10\frac{1}{3}$ **11.** $a = 1\frac{1}{4}$ **12.** $x = \frac{1}{2}$ **13.** $y = \frac{4}{15}$ **14.** $x = -\frac{3}{7}$ **15.** $y = 1\frac{2}{9}$ **16.** $x = 10\frac{17}{24}$ **17.** $y = 4\frac{1}{2}$ **18.** $\frac{1}{4}$ **19.** 4 **20.** 9 **21.** 5 **22.** 19 **23.** 12 **24.** 9 **25.** 0 **26.** –6 **27.** 11 **28.** 2 **29.** 18 **30.** –1

DAY 25

1. $(7x + 2)(x + 3)$ **2.** $(2x - 7)(x - 2)$ **3.** $(4x - 3)(x + 5)$ **4.** $(5x - 2)(x + 3)$ **5.** $(3x + 1)(x + 4)$ **6.** $(5x - 4)(x + 2)$ **7.** $(4y - 3)(y - 2)$ **8.** $(3x - 5)(3x - 2)$ **9.** $(3x + 4)(2x - 1)$ **10.** $(2n - 5)(3n + 2)$ **11.** $(4x - 7)(x + 1)$ **12.** $(3a - 2)(3a - 4)$ **13.** $(5x - 4)(2x - 3)$ **14.** $(3x - 1)(2x + 1)$ **15.** $(x + 2)(2x + 3)$ **16.** $(3x - 1)(4x + 5)$ **17.** $(5x - 1)(4x + 3)$ **18.** $(5x + 2)(3x - 1)$ **19.** $(3x - 4)(2x - 3)$

DAY 27

1. $x = 1$ **2.** $x = 8$ **3.** $x = 3$ **4.** $x = 24$ **5.** $y = -20$ **6.** $y = 6$ **7.** $y = 15$ **8.** $y = 24$ **9.** $x = 12$ **10.** $x = 3$ **11.** $x = 10$ **12.** $x = 2$ **13.** $b = 1$ **14.** $m = -2$ **15.** $x = 15$ $(x \neq 0)$ **16.** $t = -\frac{1}{3}$

DAY 28

1. 18 **2.** 39 **3.** 32 **4.** 43 **5.** 42 **6.** 29 **7.** 52 **8.** 23 **9.** $-14x + 6$ **10.** $6 - 7x$ **11.** $4x - 5$ **12.** $5x + 15$ **13.** $2y^2 - 2y - 15$ **14.** $2n^2$ **15.** $2x^2 - 10x - 8$ **16.** $12m - 12$ **17.** $-29y + 28$ **18.** $8m - 12$

DAY 29

1. $8x + y$ or $8(x) + y$ **2.** $3d + 2$ or $3(d) + 2$ **3.** $11(y - 3)$ **4.** $\frac{5}{x+4}$ **5.** $a^3 + 2^3$ or $a^3 + 8$ **6.** $12 - n^2$ **7.** $\frac{3x+2}{2-3x}$ **8.** $(5+4a)^2$ **9.** $x + 6 = 7$ **10.** $1 = 3 + s$ **11.** $-5 + |p| = 0$ **12.** $-4 > -3$ **13.** $(2)^2 + 10 < (4)^3$ **14.** $(25 \div 5) + 4^2$ or $\frac{25}{5} + 4^2$ or 21 **15.** $42 = 6 \bullet x$ or $42 = 6x$ **16.** $6^3 < 300$

DAY 30 – lcm, gcf

1. 24, 2 **2.** 10, 5 **3.** 18, 3 **4.** 12, 2 **5.** 40, 2 **6.** 24, 4 **7.** 60, 2 **8.** 30, none **9.** 12, none **10.** 20, 2

DAY 31

1. 168 **2.** 350 **3.** 336 **4.** 990 **5.** 540 **6.** 5088 **7.** 360 **8.** 4400 **9.** 168 **10.** 504 **11.** 72 **12.** 144

DAY 32

1. 14 **2.** 5 **3.** 2 **4.** 6 **5.** 12 **6.** 4 **7.** 24 **8.** 6 **9.** 6 **10.** 6 **11.** 18 **12.** 6 **13.** 6 **14.** 12

DAY 33

1. 6 **2.** 13 **3.** 8 **4.** 37 **5.** 30 **6.** 24 **7.** 169 **8.** 13 **9.** 29 **10.** 121 **11.** 12 **12.** 78

DAY 34 – gcf, lcm

1. 14, 168 **2.** 6, 990 **3.** 5, 350 **4.** 12, 540 **5.** 2, 336 **6.** 4, 5088 **7.** 24, 360 **8.** 20, 4400 **9.** 144 **10.** 54 **11.** 770 **12.** 663 **13.** 72 **14.** 320 **15.** $12x^3y^2$ **16.** 550

DAY 35 – gcf, lcm

1. 24, 720 **2.** 7, 1470 **3.** 22, 264 **4.** none, 300 **5.** 12, 504 **6.** 5, 720 **7.** 10, 2640 **8.** 5, 8460 **9.** 3, 72 **10.** 10, 80 **11.** 6, 240 **12.** 12, 120 **13.** 7, 70 **14.** 6, 72 **15.** 4, 360 **16.** 15, 300

DAY 36

1. 15 **2.** 9 **3.** 3 **4.** 12 **5.** 60 **6.** 16 **7.** 12 **8.** 9 **9.** 12 **10.** 8 **11.** 6 **12.** 8 **13.** 7

DAY 37

1. a. $6(x + 2)(x - 3)$ **b.** $x^3(x + 6)(x - 4)$ **2.** 1 and 11 **3.** The factor of 9 should be factored out. **4.** Answer will vary. Find two numbers whose sum is -9 and whose product is 20. **5.** $(x + 7)(x - 8)$ **6.** 3 and 8 **7. a.** $(x+12)(x-10)$ **b.** $3x(x - 2y)(x+5y)$ **c.** $2(x - 2)(x - 3)$ **8.** The numbers "a" and "b" must have the same sign (both positive **or** both negative). **9. a.** $(x + 11y)(x - 9y)$ **b.** $5x(x - 3)(x+5)$ **c.** prime **10.** The factorization yields a middle term of $-2x$ rather than $+2x$. The correct factorization is $(x + 5)(x - 3)$. **11. a.** $(x + 4)(x - 12)$ **b.** $(x + 15)(x - 1)$ **12. a.** $(x - 4)$ **b.** $(x + 5)$ **13.** $(x + 5y)(x - 2y)$ **14.** c **15.** b **16.** a **17.** d **18.** b

DAY 38

3. b. 55 squares **3. c.** 30 squares

DAY 39

1. $a < 6$ **2.** $b > 12$ **3.** $c \leq 20$ **4.** $3 < d < 5$ **5.** $e \leq 13$ **6.** $5 < f < 8$ **7.** $g \geq 9$ **8.** $6 < h < 12$ **9.** $j \geq 3$ **10.** $k \leq 19$ **11.** $l \geq 15$ **12.** $m \geq 40$ **13.** $2 < n < 7$ **14.** $p \geq 23$ **15.** $10 < q < 15$ **16.** $r \leq 52$ **17.** $7 < s < 15$ **18.** $t \geq 33$ **19.** $u \geq 17$ **20.** $v \leq 29$ **21.** $w \leq 18$ **22.** $5 < x < 9$ **23.** $y \geq 47$ **24.** $z \leq 27$

DAY 40

1. $(u + 6v)(u - 9v)$ **2.** $(u + 2v)(u - 7v)$ **3.** $(st - 13)(st - 2)$ **4.** $(xy + 3)(xy + 4)$ **5.** $(10 - m)(m + 4)$ or $(m - 10)(-m - 4)$ **6.** $(x + 7y)(x - 4y)$ **7.** $(x + 6)(x - 7)$ **8.** $(x + 16y)(x - 12y)$ **9.** $(u + 5v)(u - 9v)$ **10.** $(u + 3v)(u - 5v)$ **11.** $(3x + 2)(4x + 3)$ **12.** $(3y + 4)(5y + 2)$ **13.** $(3z - 4)(4z + 3)$ **14.** $(2z + 3)(4z - 3)$ **15.** $(8mn - 5)(3mn - 8)$ **16.** $(4m + 7n)(7m + 8n)$ **17.** $(3x + 4)(5x + 2)$ **18.** $(2y + 3)(3y + 4)$ **19.** $(4z - 3)(3z + 4)$ **20.** $(4z + 3)(5z - 2)$

DAY 41

1. $m = 2, b = 3$

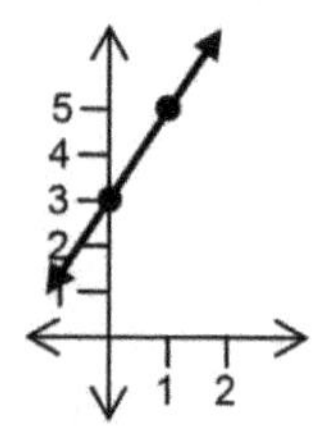

2. $m = -3, b = 1$

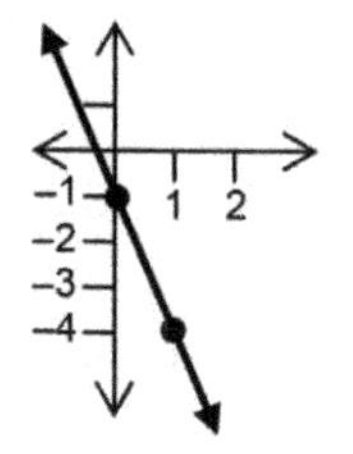

3. yes **4.** no **5.** yes **6.** no

DAY 42

1. $3x - y = -1$ **2.** $3x - 4y = -16$ **3.** $x - 2y = -4$ **4.** $22x + 18y = 17$ **5.** $3x + 9y = 22$ **6.** $4x + y = 3$ **7.** $2x + 3y = -1$ **8.** $x + y = 1$ **9.** $y = 5$ **10.** $-4x + y = 5$ **11.** $-x + y = 18$ **12.** $y = 7$

DAY 43

1. 3 **2.** 0 **3.** undefined **4.** $y = -5$ **5.** $x = -8$

DAY 44

1. parallel **2.** parallel **3.** neither **4.** perpendicular **5.** neither **6.** neither **7.** perpendicular **8.** $3x - 4y = 18$ **9.** $3x - 2y = -11$ **10.** $2y = -1$ **11.** $3x - y = -6$ **12.** $-x + y = 4$ **13.** $x = 3$ **14.** $-2x + 3y = 27$

CPSIA information can be obtained
at www.ICGtesting.com
Printed in the USA
FSHW022339310720
72260FS